THE ORIGIN AND EVOLUTION OF LIFE ON EARTH

THE MAGILL BIBLIOGRAPHIES

Other Magill Bibliographies:

The Atomic Bomb—Hans G. Graetzer and Larry M. Browning
Cycles in Humans and Nature—John T. Burns
Energy—Joseph R. Rudolph, Jr.
Environmental Studies—Diane M. Fortner
The History of Science—Gordon L. Miller
Human Genetics—Jean Helgeson
The Immigrant Experience—Paul D. Mageli
Native Americans—Frederick E. Hoxie and Harvey Markowitz
Pan-Africanism—Michael W. Williams
The Vietnam War in Literature—Philip K. Jason

THE ORIGIN AND EVOLUTION OF LIFE ON EARTH

An Annotated Bibliography

David W. Hollar

Rockingham Community College

SALEM PRESS

Pasadena, California Englewood Cliffs, New Jersey

∞ The paper used in these volumes conforms to the American National Standard for Permanence of Paper for Printed Library Materials, Z39.48-1984.

Library of Congress Cataloging-in-Publication Data

Hollar, David W.
The origin and evolution of life on earth / David W. Hollar, Jr.
p. cm.—(Magill bibliographies)
Includes bibliographic references and index
ISBN 0-89356-683-7
1. Evolution (Biology)—Bibliography. 2. Life—Origin—Bibliography. I. Title. II. Series.
Z5322.E9H65 1992
[QH366.2] 92-12899
575—dc20 CIP

To my parents, David Wason Hollar and Roselyn Sizemore Hollar, and to the memory of my grandparents: Denson Monroe Sizemore, Gertha Holcomb Sizemore, Jecholiah Dinkins Hollar, and Arthur J. Hollar.

EDITORIAL STAFF

Contents

ACKNOWLEDGMENTS

I can thank many individuals for their assistance and patience in my completion of this work. I thank John Wilson and the staff at Salem Press for their tremendous help. I thank my parents along with Beverly, Edward, and John Stafford Merritt. Special thanks to Virginia Paige Dean for her tremendous support. I thank my many colleagues at Rockingham Community College in Wentworth, North Carolina, including John Crutchfield, Kenneth Capps, Robert Futrell, Ed McLester, Robert Maynard, Lefty Crowson, Susan Clark, T and Lindley Butler, Reid Prillaman, Dee Stanfield, William Knight, Dean Burgess, and N. Jerry Owens. I thank hundreds of outstanding students, including John Robinson, Tim Stephens, James Joyce, Fred Moore, Jeri Nykamp, Raymond Broadhead, Nancy Barnes, and Earl Armstrong.

I have been inspired by past and present collaborations with several scholars, including Gary Tally, Logene Hood, Ralph Earle, Steve Beste, Paul Asman, Edward G. Barry, Larry and Philomena Grulick, Bert Goldman, and George Seielstad. I also thank James "Mac" Robertson, Fred Vaughn, the Reverend Dale Martin, the Reverend J. C. and Helena Shore, Rodney Stewart, Julia Bowling, Todd Lemons, Ellen Hundley, Mark Evans, Leslie Evans, Doug Brinkley, Steve Hoggard, and many others.

I located many of the volumes for this annotated bibliography in the Learning Resources Center at Rockingham Community College (special thanks to Peggy Markham and Robert Foeller); the Walter C. Jackson Library at the University of North Carolina at Greensboro; the Biology, Mathematics/Physics, and Health Sciences libraries at the University of North Carolina at Chapel Hill; the Coy C. Carpenter Library at the Bowman Gray School of Medicine, Wake Forest University, Winston-Salem, North Carolina; the Yadkin County (North Carolina) Public Library; and the Rockingham County (North Carolina) Public Library.

INTRODUCTION

The question of how life originated on earth has been the subject of much speculation for all of recorded history. It continues to baffle today's scientists, although modern science has yielded many clues concerning the first living molecules and cells. Likewise, the principal theory that describes the development of life on earth, evolution, has generated a storm of controversy. Nevertheless, the theory of evolution has survived repeated scientific tests and has emerged as a pervading force not only in biology but also in other scientific fields.

The theory of evolution was first proposed during the late 1700's by the scientists Erasmus Darwin of England and Compte Georges de Buffon of France. They proposed that organisms evolve from old, now-extinct species into new species by the inheritance of acquired characteristics. According to this viewpoint, organisms could develop new abilities during their lifetimes and subsequently pass these abilities to their offspring. This mechanism of evolution was championed by the famous French naturalist Jean Baptiste de Lamarck during the early 1800's. Nevertheless, evolution by the inheritance of acquired characteristics was disproved by several scientists, including the French paleontologist/economist Georges Cuvier.

In 1858, the theory of evolution was revived with a plausible mechanism for its occurrence: natural selection. Evolution by natural selection was independently codiscovered by two English naturalists: Charles Darwin (the grandson of Erasmus Darwin) and Alfred Russel Wallace. Darwin developed his theory of evolution during a five-year (1831-1836) voyage around the world while serving as naturalist aboard the survey ship HMS *Beagle*. Darwin was struck by the similarities and differences between species from all over the world. Much of the impetus for his theory came from the diversity of species found upon the Galapagos Islands located off the western coast of South America. Upon returning to England, he published his collected data and journals from the voyage. He continued his research into species classification, animal and plant husbandry, and organismal physiology while publishing numerous research articles that subtly inferred evolutionary ideas.

During the late 1840's and 1850's, Alfred Russel Wallace studied animal and plant life in the Amazon River Basin and in the East Indies. He derived conclusions almost identical to those of Darwin and proceeded to publish his results in 1858. Aware of Wallace's effort, Darwin prepared a companion paper. The two research reports, both advancing the theory of evolution by natural selection, appeared in the July 1, 1858, issue of the *Journal of the Linnaen Society*. In 1859, Darwin followed this initial report with his revolutionary book *On the Origin of Species by Means of Natural Selection: Or, The Preservation of Favoured Races in the Struggle for Life*.

Both men had been influenced by their scientific backgrounds, by exposure to exotic life and lands, and by the early evolutionary theories of Erasmus Darwin, Buffon, Lamarck, and Cuvier. A major contribution to the Darwinian-Wallace Theory of Evolution came from the English economist Thomas Robert Malthus, who in 1798 had developed a simple, yet profound, theory relating population growth to available resources. Malthus discovered that plant, animal, and human populations increase exponentially, or geometrically, in a doubling pattern (such as 1:2:4:8:16:-32:64:128: . . . and so forth). The food supply and natural resources available to a population, however, increase arithmetically in a linear, stepwise progression (such as 1:2:3:4:5:6:7:8: . . . and so forth). Consequently, any population eventually will outstrip its resources. Large numbers of individuals in the population will starve and die, resulting in a population crash. Alternatively, the population size will level off at a stable level that does not exceed available resources.

In the Darwinian-Wallace Theory of Evolution, this Malthusian doctrine illustrates the endless competition between individuals of the same and/or different species. Within any given environment, organisms compete with one another for available food, air, water, light, and territory. The species whose individuals possess the best characteristics for dealing with harsh environments and competitors succeeds and multiplies, thereby increasing in population size. Ill-adapted species find fewer resources, are victims of predators, leave fewer offspring, and gradually wither into extinction over many generations.

Neither Darwin nor Wallace understood how new characteristics evolved in individuals of a given species, principally because the particulate inheritance of traits (that is, genes) was not discovered until 1866 and 1902. In 1866, the Austrian monk Gregor Mendel discovered the principles behind which characteristic traits are inherited in sexually reproducing organisms. In 1902, Walter Sutton and Theodor Boveri independently codiscovered that the genes that confer inherited traits are physically located on chromosomes. The work of later geneticists (such as Thomas Hunt Morgan, George Beadle, Edward Tatum, and Theodosius Dobzhansky) demonstrated that chemical- and/or radiation-induced mutations alter genes, thereby producing new characteristics in the affected offspring.

Consequently, evolution occurs when mutations in a gene or genes confer new characteristics upon a group of individuals of a given species, characteristics that may help or hinder these individuals in adapting to changing environmental conditions. If the mutated gene confers characteristics that are beneficial to these individuals, then these individuals will be better adapted to their particular environment and will outcompete other species and nonmutated individuals of the same species for available resources, thereby increasing the number of their (these individuals') offspring carrying the mutated gene. Natural selection favors certain organismal characteristics over others, depending upon the environment. For example, albinism (that is, the absence of skin pigmentation) is an advantageous camouflage for squirrels living in cold, snowy climates; however, albinism is disadvantageous for squirrels living in warm climates.

The tremendous variety of organisms worldwide, numbering in the tens of millions of different species, plus the diversity of characteristics within a given species (such as iris pigmentation and blood types in humans), are the result of billions of mutations that have occurred in living organisms since the beginning of life on earth. Geological data place the earth's age at approximately 4.6 billion years. The oldest discovered fossils are those of bacteria found within Australian stromatolite rock formations dated at 3.5 billion years. Therefore, the first life (that is, first cells) originated on earth at least 3.5 billion years ago when it was still accumulating mass due to collisions with other solar system bodies.

Much speculation has been devoted as to how the first cell, the progenote, developed. Even the simplest cells can be very elaborate in their internal structures. The basic components of a cell include four essential molecules: (1) an information molecule (such as DNA or RNA), which encodes and directs the activities of the cell, (2) a catalytic molecule (such as protein), which performs essential chemical reactions within the cell, (3) an energy-storage molecule (such as sugar or carbohydrates), and (4) a membrane-producing molecule (such as lipids), which gives the cell its protective outer barrier and which forms specialized compartments inside the cell. How these four molecular components got together in the right place at the right time with the correct DNA/RNA coding sequence to produce the first cell remains a mystery. No experiments have come close to generating life in the laboratory.

Nevertheless, theoretical and experimental work have made several scenarios for events leading up to the formation of the first cells. In the 1920's, the Russian biochemist Alexander I. Oparin proposed that the interaction of energy (such as solar radiation or lightning) with the molecules of the primordial earth atmosphere (such as nitrogen, ammonia, methane, carbon dioxide, or hydrogen cyanide) produced organic compounds such as amino acids and sugars. These organic compounds rained down upon the earth's oceans. Eventually, the oceans became a "primordial soup" teeming with these organic compounds that are essential for living cells to form.

The plausibility of the Oparin primordial soup hypothesis was verified in the 1950's during a famous experiment performed by Stanley Miller and Harold Urey. There is no predominant theory, however, describing how the components of the primordial soup combined to produce the first cell. Recent research has shifted the view of the first life-form from the cellular level to the molecular level. The organic chemicals of the primordial soup may have combined further under special environmental circumstances to generate living molecules. Autocatalytic RNA, discovered in the 1980's by Thomas Cech and Sidney Altman, has been suggested as the progenitor, or first, living molecule. Other scientists, including A. G. Cairns-Smith, believe that clay crystals may have served as patterns for the first life-forms. Regardless of the mechanism, these progenitor molecules mutated and evolved into the first cells more than 3.5 billion years ago.

Still other scientists are not convinced with these origin-of-life scenarios. These scientists argue that the odds against life forming out of a chemical soup would be astronomical, even impossible, in all the immense universe. Directed panspermia, an idea initially advanced by the famous Swedish chemist Svante Arrhenius and recently revived by Leslie Orgel and Nobel laureate Francis Crick, maintains that life was seeded on earth from outer space, followed by evolutionary change.

Unfortunately, directed panspermia does not address how life originated elsewhere.

Creationists dispute the primordial origin of life and the evolution of life on earth. They maintain that life was created in present form by a supernatural being (such as God) and that humanity is the center of the universe. Some creationists place the time of creation of the earth, the universe, and life to approximately six thousand years ago. Creationists base their theory upon the Genesis creation account in the Bible and upon faults in evolutionary data. Unlike evolutionists, however, they do not supply any geological, paleontological, anatomical, genetic, or biochemical evidence to support their theory.

Much of the dissension between creationists and evolutionists revolves around humankind's relationship to "lower" animals and upon the existence of God. With respect to the latter, scientists from both groups are guilty of believing that evolution refutes the existence of God. It does not, nor did Darwin and Wallace intend for it to do so. With respect to humanity's relationship to "lower" animals, evolution has removed humankind from the center of the universe and has displaced it from special status above other organisms.

This bibliography addresses the origin and evolution of life on earth. It is aimed at anyone interested in the subject, regardless of whether the reader is a layperson, a high school or college student, or a research scientist. The annotated references included herein encompass general works on the subject, historical surveys of evolutionary theory, theoretical discussions, detailed experimental work, and controversies/debates. This bibliography encompasses major evolutionary works from the late 1800's to the present. As such, it represents an important reference source for the serious student of evolution. The chapters are arranged both topically and sequentially in lieu of the evolutionary and geological timetables. Books referenced for each chapter are selected based upon their appropriateness for the chapter's subject matter; both general and more detailed works are included. Thousands of excellent evolution books have been written; the eight-hundred-plus works that are cited in this volume should represent a thorough sampling of the available literature, both old and new, with the addition of some nontraditional evolutionary and creationist works. It is unfortunate that not every evolutionary work can be cited. The bibliography should also be an important source of information for the beginning student of biology, the evolutionary scientist, and the creation scientist.

Chapter 1 ("General Studies") identifies major survey works on evolution and evolutionary theory. These works include Darwin's preeminent *On the Origin of Species* as well as popular works by prominent twentieth-century evolutionary scientists, philosophers, and writers. Chapter 2 ("Evidence Supporting Evolution") discusses evolution using evidence from several major fields, including paleontology, ecology, and biochemistry. Paleontology includes the fossil, archaeological, and geological studies of organismal remains. Ecology considers the interactions and spatial distributions of organisms and species (that is, where organisms live, and why). Biochemistry provides the strongest support for evolution with interspecies comparisons of protein amino acid sequences, DNA (that is, deoxyribonucleic acid) nucleotide sequences, and physiological processes. Chapter 3 ("Evolutionary Controversies") focuses on debates concerning the mechanisms of evolution as well as alternate theories describing the origin and evolution of life on Earth, including creationism.

Chapter 4 ("Cosmology") and chapter 5 ("Primordial Life") consider works devoted to presumed conditions preceding the formation of life. Many of these works revolve around the anthropic principle, the premise that life is a natural development of the evolution of the universe. Such works examine the big bang theory of the universe's origin, the formation of elements, molecules, stars, and galaxies, and the possible fates of the universe (such as black holes, dissipation, or collapse). Chemical and physical processes essential for life are described in these books, including Oparin's primordial soup, the Miller-Urey experiment, and the evolution of the first living molecules and cells. Works describing the genetics and biochemistry of mutation, the initial trigger of evolutionary change, are presented as well.

Chapter 6 ("The Prokaryotes") presents works describing bacteria, the single-celled organisms that dominated the first 2.5 billion years of life on Earth and that continue to be extremely prevalent today. While being microscopic, they constitute much of the total earth biomass. In many ways, their evolutionary development has made possible the evolution of higher multicellular life. Bacteria modified the early earth, particularly the atmosphere, with the evolution of cellular respiration, photosynthesis, and free atmospheric oxygen. Mobile genetic elements (that is, DNA) within bacterial cells may have become the parasitic non-cellular living molecules known as viruses (chapter 7).

From prokaryotes evolved the multicellular eukaryotes (chapter 8). Works on eukaryotic evolution describe the evolution of protists, fungi, plants, and animals. Major evolutionary events in the development of eukaryotes include eukaryotic origins from symbiotic bacteria, greater complexity of cells, the ability of cells to work together as larger units (such as tissues, organs, or multicellular organisms), and the ability of cells to differentiate and specialize to perform specific tasks within the multicellular organism. Comparative anatomy has proven to be very useful in the evolutionary classification of eukaryotes.

Chapters 9 through 14 deal with special avenues of evolutionary thought. Evolution has spawned considerable debate (chapter 9) concerning the relatedness of various species or families of species. Evolutionists attempt to construct family trees called phylogenies to illustrate the evolutionary descent of various species. Furthermore, the geological record indicates the occurrence of periodic mass extinctions in which large numbers of species died, possibly due to asteroid/cometary impacts or other catastrophes (chapter 9). One extinction, the Cretaceous, ended the dominion of reptiles, including the dinosaurs, on Earth approximately sixty-five million years ago. Some dinosaur descendants, however, may have survived: the birds (chapter 10). The decline of reptiles precipitated the rise of mammals, furry creatures that have strong social bonding and that feed their young with milk. Strong mammalian social bonding led to the evolution of intelligence in two mammalian groups (chapter 11): the cetaceans (such as whales or dolphins) and the primates (such as humans, chimpanzees, or gorillas). Several human species have existed during the past six million years (chapter 12), of which we (that is, *Homo sapiens sapiens*) are the only current representative. Humans have remarkable evolutionary potential both on and off the earth. Furthermore, humans have the ability to contemplate their own origins and existence as well as the ability to seek life that has evolved independently elsewhere in the universe (chapter 13).

Evolution represents not only the strongest scientific theory describing the development of life on Earth but also a beautiful model interconnecting all living things. It provides us with an almost-spiritual oneness with all living organisms and with the universe, a oneness that is emphasized in many human religions. Nevertheless, it has been grotesquely misinterpreted by many people, including politicians, economists, and scientists, as justification for the subjugation and exploitation of so-called inferior peoples. We humans have the potential to control our own evolution for the betterment of all people and for all living things.

Chapter 1
GENERAL STUDIES

Asimov, Isaac. *Beginnings*. New York: Walker, 1987.
Asimov, the biochemist, science-fiction author, and popularizer of science, has done it again, this time assembling a survey of evolution specifically oriented to the layperson. The book clearly and concisely describes the origin and evolution of life on earth, including major events that contributed to the evolution of multicellular eukaryotes. The evolution of intelligence in mammalian species, including human evolution, is described. Complex topics are presented with great simplicity. The book is an excellent starting point for the student of evolution.

Attenborough, David. *Life on Earth: A Natural History*. Boston: Little, Brown, 1979.
Attenborough's book, derived from his popular PBS series of the same name, is a clear, dynamic discussion of life and its development on earth. He proceeds sequentially through the evolutionary time scale, moving from bacteria and protists in the early chapters to plants and animals later. Most of this thirteen-chapter work is devoted to animals, particularly the vertebrates (such as fish, amphibians, reptiles, birds, and mammals). The primates are studied in depth. The book includes beautiful photographs from the television series.

____________. *The Living Planet*. Boston: Little, Brown, 1986.
This concise, beautifully illustrated book is designed to accompany Attenborough's popular PBS television series on natural history and evolution. Attenborough presents his usual unique and lively discussion of the history of life on earth, including the evolution of major species groups and the intricate interconnectedness between all living organisms. The book is very easy reading for the general public. The examples used to convey key concepts are extremely well chosen. This work is an exciting introduction to evolution, natural history, and humanity's impact upon the earth.

Avers, Charlotte J. *Process and Pattern in Evolution*. New York: Oxford University Press, 1989.
An outstandingly comprehensive survey of evolutionary theory. Avers deals with the mechanisms of evolutionary change, the molecular basis of life, the origins of life on earth more than 3.8 billion years ago, the evolutionary development of life on earth, human evolution, trends in evolutionary change, and the history of evolutionary theory and research. She cites a tremendous quantity of data and cites hundreds of references, thereby making her book an authoritative guidebook to the subject of evolution.

Ayala, Francisco, and James W. Valentine. *Evolving: The Theory and Processes of Organic Evolution*. Menlo Park, Calif.: Benjamin/Cummings, 1979.
Ayala and Valentine, two prominent evolutionary geneticists, describe the theory of evolution, major concepts, and the various studied mechanisms of evolution. Major evolutionary events from the origin of life to the present are included. The genetic basis of evolutionary change and the biology of species populations are stressed throughout the book. The book is extremely well written and is very thorough in its coverage of the subject. Numerous examples, illustrations, and data tables supplement the text, along with extensive reference lists for further study.

Barrett, Paul H., et al., eds. *Charles Darwin's Notebooks: 1836-1844*. Ithaca, N.Y.: Cornell University Press, 1987.
This lengthy collection of comments, philosophical speculations, scientific research, and personal opinions of Charles Darwin beginning with his return to England after the *Beagle*'s historic voyage provides the reader with a unique glimpse at the scientist's thoughts during his formulation of evolutionary theory. Darwin discusses many topics, including the process of speciation, methods of classifying organisms, geology, the fossil record, the works of Cuvier and Hume, and his opinions of various other scientists' work. Barrett and colleagues provide important reference notes to assist the reader.

Barrett, Paul H., and R. B. Freeman, eds. *The Works of Charles Darwin*. Vols. 1-29. New York: New York University Press, 1987.
All Charles Darwin's writings are found in this valuable collection, which numbers in the thousands of pages. Darwin's *Origin of the Species*, *The Descent of Man*, naturalist's notes and diary aboard the *Beagle*, and research notebooks throughout his life are found herein. Barrett and Freeman provide important background material, including extensive biographies of such personalities as Charles Darwin and Erasmus Darwin, and useful indexing for the reader. This collection is the most complete survey of Darwin's thinking on numerous subjects, the foremost of which is, of course, evolution.

Bonner, John Tyler. *The Evolution of Complexity by Means of Natural Selection*. Princeton, N.J.: Princeton University Press, 1988.
This discussion of evolution is targeted for the general public, although it would be of interest to evolutionary scientists as well. It describes the evolution of communities of species and uniquely compares such evolution to the cellular development and differentiation that occur within an individual multicellular organism such as ourselves. In so doing, Bonner illustrates that cells have evolved to great levels of complexity during the history of life on earth.

Bowler, Peter J. *Evolution: The History of an Idea*. Berkeley: University of California Press, 1984.

Bowler's book is a thorough account of evolutionary thought. He discusses the mythological and religious views of life from ancient times to the present, the beginnings of evolutionary ideas in the 1700's, and modern evolutionary theory from Darwin and Wallace to today. He details the processes and patterns of evolutionary change very succinctly. He also addresses problems in evolutionary theory, raging debates, and challenges from creationists. The book provides a comprehensive but clear introduction to evolution for the layperson.

Brandon, Robert N. *Adaptation and Environment*. Princeton, N.J.: Princeton University Press, 1990.

Brandon, a philosopher, defines and describes the mechanisms of evolution. Evolution is the ability of organisms to leave more offspring based upon the environmental adaptability of their genes. He proposes an empirical argument for distinguishing between environmental natural selection and other evolutionary factors such as genetic drift. He also challenges certain evolutionary assumptions.

Brooks, Daniel R., and E. O. Wiley. *Evolution as Entropy: Toward a Unified Theory of Biology*. Chicago: University of Chicago Press, 1988.

This concise book is a discussion of evolution and the chemical/physical processes that drive the process within living organisms on earth. Brooks and Wiley confront a major topic in evolution and physics: entropy. Entropy is increasing disorder in all physical processes occurring everywhere throughout the universe. Evolution is an orderly process that appears to counteract entropy. The authors relate evolution and entropy within the complex energy systems and food webs that tie together all living organisms on earth. Excellent illustrations and references are provided to back up key points.

Bylinsky, Gene. *Life in Darwin's Universe: Evolution and the Cosmos*. Garden City, N.Y.: Doubleday, 1981.

Bylinsky's interesting book is an entertaining survey of the evolution of the universe from the big bang theory to the domination of the earth by humans. In species evolution, he looks at various lines of evidence that demonstrate the relatedness of species, including the comparative anatomy of eyes between humans, squid, and insects. He examines chance twists in the evolution of life and speculates what life might be like on other worlds. Chapter 8, "Future Man," and chapter 9, "Planet Search," explore humanity's future evolution on and off the earth.

Crawford, Michael, and David Marsh. *The Driving Force: Food, Evolution, and the Future*. New York: Harper & Row, 1989.

This brilliantly written book is a unique linkage between evolution and the nutritional requirements of organisms. It summarizes the history of life on earth in layperson's terms and then proceeds to focus upon the basic organismal requirements and physical processes that cause evolution to occur. Organisms on earth are arranged in closely interconnected food webs. Competition for available resources is a driving force in the evolution of new species. Crawford and Marsh consider the colonization of land by plants and animals, the efficiencies of predators, the evolution of intelligence, and human evolution in light of food/energy requirements. They also speculate on the future of human evolution with respect to factors such as heart disease and stress.

Darwin, Charles. *On the Origin of Species by Means of Natural Selection: Or, The Preservation of Favoured Races in the Struggle for Life*. London: J. Murray, 1859. Reprint. New York: Oxford University Press, 1968.
Without question, Darwin's monumental work, the first book on evolution, is the definitive source on the subject. Darwin convincingly argues for the evolution of living organisms via natural selection. He supports his arguments with extensive data from his own and other scientific researches, including those of Alfred Russel Wallace. Much of his own work came from the observation of insect societies. Darwin's book is pleasant and informative reading, although it does contain some errors, to which the author admits due to the state of scientific knowledge at its initial publication in 1859.

____________. *"The Origin of Species" and "The Descent of Man."* New York: Modern Library, 1977.
This volume is a compilation of Darwin's two greatest works. In *On the Origin of Species*, he expounds his revolutionary new theory of evolution. In the sequel, *The Descent of Man*, Darwin refines his view of evolution by proposing more detailed mechanisms by which the evolutionary process occurs. Also, he clearly interprets humans as being animals and as being probable relatives of primates, a position that he had not previously dared to advance. He describes the evolution of physical characteristics for males and females of various species, as well as sexual selection for specific traits. He describes the inheritance of characteristics at corresponding periods of life in parents and their offspring as well. *The Descent of Man* is mostly behavioral in scope, and it includes examples from insect, fish, reptile, bird, and mammalian species.

Dawkins, Richard. *The Blind Watchmaker*. New York: W. W. Norton, 1986.
Dawkin's book is a general introduction to evolution for the layperson. He describes the physical and chemical principles behind evolution with unique explanations, examples, and analogies. Special emphasis is placed upon gradual change and the element of chance in the evolution of new species. He presents evidence supporting the rapid punctuated equilibrium (catastrophic) view of evolution in which many species become extinct in a short time frame, followed by an explosion of new species evolving from the survivors.

Dobzhansky, Theodosius. *Evolution, Genetics, and Man*. New York: John Wiley & Sons, 1955.
Dobzhansky's book is the traditional introduction to evolution, although his style and command of relevant supporting data make the book a valuable reference. He approaches evolution strongly from the genetic viewpoint, which is very appropriate. Gregor Mendel's principles of inheritance are described along with the interrelationship of organismal genotype and phenotype. Dobzhansky also looks at the evolution of multicellular organismal development and differentiation. He discusses the evolution of sex and compares vitalistic (spiritual) and mechanistic (evolutionary) views of life.

__________. *Genetics and the Origin of Species*. New York: Columbia University Press, 1937, 1982.
This classic evolutionary book revolutionized twentieth-century biology when it united genetics, population biology, and evolutionary systematics/taxonomy. The book is based upon Dobzhansky's lectures at Columbia University in 1936. Dobzhansky correctly promotes genetics as the basis of evolutionary change in that genes mutate, thereby affecting organismal traits which are naturally selected by the environment. This work ushered in the modern synthetic theory of evolution by tying together Darwinian-Wallace evolution, genetics, biochemistry, and systematics.

__________. *Mankind Evolving: The Evolution of the Human Species*. New Haven, Conn.: Yale University Press, 1962.
Dobzhansky, one of the great evolutionary biologists of the twentieth century, assembled in this book a storehouse of data from diverse scientific disciplines to support modern ideas concerning evolution. The book is entertaining and approaches evolution from historical, philosophical, and theoretical avenues. His work focuses primarily upon the evolution of humankind, looking at such evolutionary factors as genes for intelligence, body form, and blood type. He supports his arguments with objective data, including identical twin studies. He uniquely addresses the psychological and cultural evolution of human races and classes, as well as humanity's future.

__________, Francisco Ayala, James W. Valentine, and W. Davis. *Evolution*. San Francisco: W. H. Freeman, 1977.
This excellent introduction to evolutionary theory and mechanisms is one of the best, most up-to-date works in the field. It taps a wealth of scientific data to illustrate key concepts including mutation, chromosome abnormalities, selection pressures, genetic drift, adaptive radiation, and molecular clocks. Numerous examples from a wide variety of species, including the authors' own fruit fly *Drosophila melanogaster*, are provided. These four authors are among the world's

top evolutionary scientists. Their ideas are presented with great clarity backed up by excellent data tables, illustrations, and reference lists.

Dodson, Edward O. *Evolution: Process and Product*. New York: Reinhold, 1960.
This book is another traditional introduction to evolution. It discusses the concept of species, the classification of species (taxonomy and systematics), the history of evolutionary theory, and the origin of life. The evolution of cells, organisms, and species are discussed within the context of genetics (DNA and mutation). Mutation, natural selection, and population biology are described in detail to illustrate the mechanisms of evolution. The comparative anatomical and physiological analysis of different species also is provided as evidence supporting the occurrence of evolution.

Edey, Maitland A., and Donald C. Johanson. *Blueprints: Solving the Mystery of Evolution*. Boston: Little, Brown, 1989.
Edey and Johanson's comprehensive analysis of evolutionary theory and its history spans from the founding scientists in the late eighteenth century to the elaborate molecular evolutionary studies of today. Chapter 1, "Six Who Helped Lay the Groundwork," describes the work of Erasmus Darwin, Jean Baptiste de Lamarck, and others who paved the way for Darwin and Wallace to refine and expound their theory of natural selection. Darwin's experiences aboard the *Beagle*, publication of *On the Origin of Species*, and public reaction to evolution are described in detail. Edey and Johanson incorporate the important contributions of Gregor Mendel, Thomas Hunt Morgan, Oswald Avery, and other twentieth-century geneticists to the advance of evolutionary theory. They also discuss new views on evolution such as adjustments in organismal classification and patterns of evolution advanced by Carl Woese of the University of Illinois, Champaign-Urbana.

Ehrensvard, Gosta. *Life, Origin and Development*. Chicago: University of Chicago Press, 1960.
This brief book is a synopsis of the early evolution of life on earth. It discusses cosmic evolution, chemical evolution on earth, and the formation of the first cells in a concise, readable format. Alexander I. Oparin's primordial soup hypothesis, in which atmospheric chemicals combined to produce key organic hydrocarbons (such as amino acids, sugars, lipids, or nucleotides) more than four billion years ago is emphasized. Ehrensvard speculates upon early mutational events in beginning cells that led to heterotrophy (feeding upon others) and autotrophy (self-feeding processes such as photosynthesis).

Eiseley, Loren. *Darwin's Century: Evolution and the Men Who Discovered It*. Garden City, N.Y.: Doubleday, 1958.
In this book, Eiseley discusses the lives and works of the scientists who stimulated

the Darwinian-Wallace Theory of Evolution. At the same time, he addresses the impact of evolution upon science and society. Eisely describes the works of Erasmus Darwin and Jean Baptiste de Lamarck, the early fossil work of paleontologists, Darwin's voyage aboard the *Beagle*, the genetics experiments of Gregor Mendel, and the anthropological studies of ancient humans and missing links. Eiseley's book is fast-paced, scholarly, and very informative for the general public.

____________. *The Immense Journey*. New York: Random House, 1946.
Anthropologist Eiseley's book is a thrilling presentation of evolution and natural history from a philosophical viewpoint. His writing is very imaginative, shifting back and forth between the natural world of today and past events in the history of life on earth. The book is a fascinating meditation upon life and its meaning. Special chapters include "The Secret of Life" and "The Dream Animal." He also considers human evolution and humanity's future.

Eldredge, Niles. *Unfinished Synthesis: Biological Hierarchies and Modern Evolutionary Thought*. New York: Oxford University Press, 1985.
This book, written by a leading evolutionary thinker, is a unique, exciting discussion of evolutionary theory for a general audience. The major concepts of evolution are discussed along with points of dispute between scientists concerning the possible mechanisms of evolutionary change. Numerous species examples and illustrations are provided. The evidence to support evolution is compelling, and the presented data provide profound insights into the thinking of modern evolutionary scientists.

Endler, John A. *Natural Selection in the Wild*. Princeton, N.J.: Princeton University Press, 1986.
Endler's book is a survey of evolutionary mechanisms and how they are studied. He describes the population biology of natural wild animal and plant species while measuring the effects of natural selection upon various traits (genes) in each population. He concludes that natural selection is an essential process in evolutionary change, although it is not the only important factor. Continued mutation, genetic drift, founder effects, geographical isolation, migration, and so forth, also play key roles in evolution and the origin of new species.

Fox, Sidney W. *The Emergence of Life: Darwinian Evolution from the Inside*. New York: Basic Books, 1988.
Fox, the eminent theoretical evolutionary scientist, presents the intricate principles behind evolutionary change. He describes the physical and chemical evolution of the universe and of life. He discusses important processes such as entropy (disorder), mutation, and speciation in a light that is quite different from the usual introduction to evolution. The emphasis in this book, as in other Fox works, is

the self-organization of living molecules, cells, and living organisms during the history of life on earth. The book opens up one's mind to new ways of looking at life.

Futuyma, Douglas J. *Evolutionary Biology*. Rev. ed. Reading, Mass.: Sinauer Associates, 1986.
Futuyma's comprehensive introduction to evolution and population biology is a serious survey of the subject intended for the biology student, although it is written with great clarity for sufficient understanding by the layperson. The book covers all aspects of evolutionary theory, evolutionary mechanisms, population genetics, and the mathematics of population dynamics, including selection pressures, species isolating mechanisms, changes in allele frequencies over time, and the Castle-Hardy-Weinberg equilibrium equations. The book is well illustrated with numerous example situations. Extensive reference lists also are provided.

Gould, Stephen Jay. *Ever Since Darwin: Reflections in Natural History*. New York: W. W. Norton, 1973.
In this book, the eminent Harvard University paleontologist Gould discusses the ideas and mechanisms behind evolution. He describes the controversies surrounding the theory plus the impact of evolutionary theory upon society. He debates the tenets of creationism and belabors how science often is misunderstood by society. He closes the book with several chapters dealing with humanity's evolutionary heritage, including the evolution of races, sociobiology, intelligence, and violence.

Grant, Verne. *The Origin of Adaptations*. New York: Columbia University Press, 1963.
In a lengthy survey of evolution, Grant discusses theories concerning the origin of life and the mechanisms of evolution. He begins with a discussion of life's origins with special emphasis placed upon Alexander I. Oparin's primordial soup hypothesis. Then he describes the characteristics and evolution of cells and multicellular organisms. Special topics include genetic drift (random changes in species' traits over time), the divergence of species, the population genetics of species, and the evolutionary pressures that select for or against specific traits.

Greene, John C. *The Death of Adam: Evolution and Its Impact on Western Thought*. Ames: Iowa State University Press, 1959.
Greene's book is an enthralling essay covering evolution, the development of the theory, and the theory's impact upon society. He describes the works of Lamarck, Darwin, Wallace, and others, and scientific evidence supporting evolution. He links the development of evolutionary theory to earlier scientific advances, including Sir Isaac Newton's organization of the universe. Special chapters

describe the origin of human races and societal reactions to evolution. The book is written specifically for the layperson.

Haldane, J. B. S. *The Causes of Evolution*. Ithaca, N.Y.: Cornell University Press, 1966.
Based upon his 1931 lectures at Aberystwyth, Haldane's book is a clear-cut description of evolution for the layperson. The eminent scientist, who cochampioned the primordial soup hypothesis with Alexander I. Oparin, describes DNA, genes, chromosomes, mutation, chromosome variations within a species, and the various mechanisms of natural selection by which the environment favors certain types of genetic material/physical characteristics over others. Haldane questions misconceptions about the concept of fitness, in which he explains that organismal adaptability does not necessarily depend upon factors such as size or strength. Haldane's book includes a simple mathematical appendix.

Hamilton, Terrell H. *Process and Pattern in Evolution*. New York: Macmillan, 1969.
Hamilton's book is a short, concise, and very clear presentation of evolutionary theory and mechanisms. He describes the science of genetics and how genes are transmitted from generation to generation under the influence of natural selection. He describes organismal adaptations and selective pressures that favor one allele of a gene over another. He also describes the genetics and ecologies of species populations plus the formation of new species by the geographical isolation of populations.

Harvey, Paul H., and Mark D. Pagel. *The Comparative Method in Evolutionary Biology*. New York: Oxford University Press, 1991.
In this informative survey of evolutionary theory and evolutionary mechanisms, scientists Harvey and Pagel explore one of the primary evidences used to support evolution: the comparative approach. Different species, both extant and extinct, are compared and contrasted in terms of their anatomies, physiologies, biochemistries, and embryologies. The authors apply these techniques to answer several challenging questions in evolutionary biology, such as the effects of a large genome (DNA supply) size upon the evolutionary development of a species.

Hoffmann, Ary A., and Peter A. Hoffmann. *Evolutionary Genetics and Environmental Stress*. New York: Oxford University Press, 1991.
The Hoffmanns explore a major idea in evolutionary biology, the concept that the rapid extinction of once-dominant species and the simultaneous explosion of new replacement species may occur during periods of severe environmental stress, leading to shifts in the natural selection favoring certain organismal traits over others. They explore the genetics and dynamics of such changes, citing several examples from the evolutionary record to support their premise. Their work has

tremendous applications to environmental conservation strategies in light of human pollution of the earth.

Hotton, Nicholas, III. *The Evidence of Evolution*. New York: American Heritage, 1968.
Hotton's book is a brief, very general survey of evolution. He discusses the history of evolutionary theory, including an account of Charles Darwin's voyages and researches. A consideration of evolutionary evidence is presented, including comparative anatomical studies between living and extinct species and evidences from the fossil record. Hotton provides brief historical sketches of major evolutionary scientists (such as Darwin and Lamarck), a geological timetable of evolutionary events over the last several hundred million years, and excellent photographs/illustrations.

Huxley, Julian. *Evolution: The Modern Synthesis*. New York: Harper & Row, 1942.
Huxley, who followed in his father's footsteps as a major advocate of the Darwinian theory of evolution, presents the case for evolution in this book, which presents the history and precepts of the theory. In a thorough, six-hundred-page survey, Huxley describes the major tenets of evolutionary theory, the basic principles of Mendelian genetics, mutation, and the mechanisms of natural selection. He describes geographical isolation in species formation, the meaning of adaptability and fitness, and major trends in the evolution of life on earth.

____________. *Evolution in Action*. New York: Harper & Brothers, 1953.
In this rather short, general treatment of evolution for the average person, Huxley describes the history of the development of Darwinian evolutionary theory and the principal natural events that contribute to the evolutionary process. He describes how evolution works, mutational events in genes, natural selection by the environment for or against certain genes, and how new species arise in nature. He also speculates upon the development of mental activity in mammals and intelligence in humans.

Huxley, Thomas Henry. *On a Piece of Chalk*. Edited by Loren Eiseley. New York: Scribner's, 1967.
This book, written by Julian Huxley's father, consists of lectures and sketches presented by the elder Huxley to the British Association for the Advancement of Science in 1868. Thomas Henry Huxley was one of the most vocal supporters of evolution after Darwin and Wallace had published their theory. Huxley presents a natural history of life on earth in which new species evolve from ancient species. He describes geological and fossil work clearly and precisely. Editor Eiseley provides comments and supporting material, including a geological timetable with major evolutionary events arranged per geological time period.

Irvine, William. *Apes, Angels, and Victorians: Darwin, Huxley, and Evolution.* Cleveland: World Publishing, 1955.

In this popular, general work, Irvine describes the history of evolutionary theory and its proponents, beginning with ancient views and ideas. Much of the book is devoted to the life works of Charles Darwin and Alfred Russel Wallace and to public reactions to their theory. The works of other nineteenth-century evolutionists, including Thomas Henry Huxley, are described as well. The political and economic ramifications of the theory are discussed.

Jastrow, Robert, gen. ed. *The Essential Darwin*. Selections and commentary by Kenneth Korey. Boston: Little, Brown, 1984.

This book contains the majority of Charles Darwin's two principal works, *On the Origin of Species* and *The Descent of Man*, along with commentaries highlighting key points and concepts. Additionally, Darwin's autobiography and an account of his experiences aboard the *Beagle* during its five-year circumnavigation of the world are included. The book concludes with a special section illustrating how modern science, especially molecular genetics, is being used to bolster Darwin's theory.

Kurten, Björn. *The Innocent Assassins: Biological Essays on Life in the Present and Distant Past*. New York: Columbia University Press, 1991.

In this simple, interesting, and often humorously written book, Kurten discusses the major ideas of evolution for the average person. He describes the methods and experiments of physics, chemistry, biology, and astronomy to demonstrate how we know what we know about the nature of the universe and of life on earth. Chapter 6, "The Clock in the Rock," describes the methods of geology and paleontology for looking into the earth's past. Chapter 11, "Eve in Africa," discusses how the molecular analysis of human mitochondrial DNA may have traced the origins of *Homo sapiens* to a single woman who lived in eastern Africa approximately two hundred thousand years ago. Chapter 3, "Tom Thumb and King Kong," looks at size limitations for organismal evolution.

McNamara, Kenneth J., ed. *Evolutionary Trends*. Tucson: University of Arizona Press, 1990.

This interesting work is a collection of scholarly essays by evolutionary scientists aimed at explaining evolution to the general public. The essays as a whole emphasize long-term trends in the process of evolutionary change on the planet. Mechanisms of natural selection and evolution are described along with the criteria for measuring and identifying speciation, the appearance of new species, and so forth. The evolution of invertebrate and vertebrate animals is a leading highlight of this important work.

Mayr, Ernst. *Animal Species and Evolution*. Cambridge, Mass.: Harvard University Press, 1963.
This enormous book is a simple, yet comprehensive, study of evolutionary theory by one of the most prominent zoologists of the twentieth century. Mayr's work deals with major problems in evolutionary theory and clearly describes the mechanisms of evolution. Mayr intends for his book to be a compendium of relevant research studies, for it is packed with sample data. An extensive bibliography accompanies the text. Among the principal topics presented include the definition of species, taxonomy and systematics for species classification, the genetics of organismal populations, humankind as an animal species, and species-isolating mechanisms in nature. The book is slanted toward animal species due to the author's background, although some plant examples are cited.

____________. *The Growth of Biological Thought: Process, Pattern, and Evolution*. Cambridge, Mass.: Harvard University Press, 1982.
This mammoth, nine-hundred-page work, written by an eminent evolutionary scientist, is another survey of biological experimentation with respect to evolution. Mayr describes the history of evolutionary thought from the ancient Greeks to Charles Darwin to modern evolutionary genetics. He describes the nature of the gene (DNA), mutation, and natural selection for or against certain mutations (alleles of genes). As usual, he provides numerous examples and data from the scientific literature, always presenting the material in a very readable, concise form.

____________. *One Long Argument: Charles Darwin and the Genesis of Modern Evolutionary Thought*. Cambridge, Mass.: Harvard University Press, 1991.
In this exhaustive analysis of evolutionary theory, the preeminent evolutionary biologist Mayr discusses the works of Charles Darwin, Darwin's philosophy, and Darwin's development of the theory of evolution. Mayr explains many nebulous questions concerning some of Darwin's evolutionary arguments and controversies surrounding Darwin's theory. This book represents a clear presentation of the Darwinian theory of evolution for the general public.

____________. *Systematics and the Origin of Species*. 1942. Reprint. New York: Columbia University Press, 1982.
Mayr, one of the top twentieth-century evolutionary scientists, propounds that evolutionary systematics (the orderly classification of organisms based upon descent) is consistent with genetics, a view enunciated first by Theodosius Dobzhansky in his classic *Genetics and the Origin of Species*. Mayr's book is equally a classic because it supplies an abundance of data verifying that natural selection acts at the level of genes within an organism. As a result, an exact physical mechanism of evolution is provided, the modern Synthetic Theory of Evolution.

____________. *Toward a New Philosophy of Biology: Observations of an Evolutionist*. Cambridge, Mass.: Harvard University Press, 1988.
This book represents a thrilling collection of provocative essays by Mayr, one of the giants of twentieth-century evolutionary biology. He philosophizes upon a number of evolutionary topics, including the importance of molecular biological data in supporting evolutionary events, the evolutionary aspects of human ethical systems, the controversy over gradualistic evolution versus punctuated equilibrium, and the search for extraterrestrial intelligence. Mayr writes with great lucidity in order to bridge the "intellectual" gap between scientists and the general public.

Milner, Richard. *The Encyclopedia of Evolution: Humanity's Search for Its Origins*. New York: Facts on File, 1990.
This collection of more than five hundred entries describes various important topics, methodologies, theories, books, scientists, and controversies in evolution. Milner, an anthropologist, beautifully orients the book toward the general public. Each entry includes a bibliography and cross-references to other entries. The book is a unique collection of biographies, bibliographies, topical surveys, and films from diverse sources, thereby making the encyclopedia both entertaining and comprehensive in scope.

Moody, Paul Amos. *Introduction to Evolution*. New York: Harper & Row, 1953.
This lengthy, well-illustrated survey of Darwinian evolutionary theory looks at the mechanisms of natural selection in natural populations of organisms and provides evidences favoring evolution from several different scientific disciplines. Evolution is presented from the comparative anatomy of various species, comparative embryology of species development, serological tests on blood and other bodily fluids from various species, and the geographic distributions of species. The genetics of species populations also is explained.

Moore, Ruth. *Evolution*. 1962. Rev. ed. New York: Time-Life Books, 1968.
This very general, pictorial account of evolutionary theory begins with a history of the development of the theory of evolution. Charles Darwin's life and work are described. The principles behind evolution are presented, including mutation of DNA and natural selection for or against new genes (traits). Evidence from the fossil record and molecular genetics (such as chromosomes and DNA) are presented. The book contains many beautiful photographs and informative diagrams that illustrate the process and history of evolution.

Ridley, Mark. *The Problems of Evolution*. New York: Oxford University Press, 1985.
Ridley's work is a clear, concise synopsis of evolutionary theory and its mechanisms. Ridley presents convincing evidence supporting evolution, although he does focus upon certain points of contention between various scientists. Such

topics include discrepancies in the fossil record, gradualistic evolution versus punctuated equilibrium (catastrophism), the Lamarckian inheritance of acquired characteristics, and microevolution versus macroevolution.

Rosen, Robert. *Life Itself: A Comprehensive Inquiry into the Nature, Origin, and Fabrication of Life*. New York: Columbia University Press, 1991.
Rosen, a Dalhousie University biophysicist, debates the concept of life and its meaning in this philosophical work. He describes the development of evolutionary theory during the past several centuries. He debates the mechanistic versus vitalistic views of life. The current prevailing mood among scientists strongly favors a mechanistic view of life, where living organisms essentially are machine-like. Rosen argues that a mechanistic interpretation of life is not essential within the context of evolution. He presents a radically alternative view of the evolutionary process that is based upon complex systems of interrelated organisms.

Sagan, Carl. *Cosmos*. New York: Random House, 1980.
Sagan's book, based upon his award-winning and internationally viewed PBS television series, is a wonderful presentation of the universe and humanity's place within the cosmos. Sagan discusses the relationships between science, scientists, and history. Special emphasis is placed upon the evolution of the universe and of life within the universe, with life on earth and elsewhere being a natural extension of cosmic evolution. The joy of discovery is felt throughout the book. Sagan's inimitable style and beautiful photography make *Cosmos* a thrilling exploration of our universe.

Seielstad, George A. *At the Heart of the Web: The Inevitable Genesis of Intelligent Life*. Boston: Harcourt Brace Jovanovich, 1989.
Seielstad, a radio astronomer with the National Radio Astronomy Observatory, has produced a beautifully concise and thought-provoking book that summarizes the evolution of the universe and of life. He describes the formation and evolution of stars and galaxies many billions of years ago, major events in the evolution of life on earth, and the future evolution of humans and the universe. He stresses the intricate interconnectedness between all living organisms on earth to produce a superorganism. He presents compelling evidence and data from many different sources to support his views. The book is very philosophical and informative, with a wealth of examples, data tables, beautiful photographs, and valuable reference lists.

Skutch, Alexander F. *Life Ascending*. Austin: University of Texas Press, 1985.
Skutch's book is a general discussion of evolution for the layperson. Skutch describes major trends in evolutionary theory, including the mutation of genes (DNA), chromosome abnormalities, and natural selection for or against these

mutated genes and abnormalities. He describes complex topics with great clarity and uses specific classic examples to illustrate key points.

Smith, John Maynard, ed. *Evolution Now: A Century After Darwin*. San Francisco: W. H. Freeman, 1982.
This book is a collection of essays on evolutionary theory and mechanisms written by a host of prominent twentieth-century biologists, including Francis Crick, Leslie Orgel, and Stephen Jay Gould. Various topics covered in this compendium include the origin of life, gradualistic evolution versus punctuated equilibrium (evolution in bursts), cladistics (evolutionary relationships), the evolution of immunity, the evolution of behavior, and Lamarckian evolution (inheritance of acquired characteristics/traits). This volume provides unique insights into the thinking of some of the world's most creative scientific minds.

Stanley, Stephen M. *Macroevolution: Pattern and Process*. San Francisco: W. H. Freeman, 1979.
In this introduction to evolution for the general public, Stanley describes the principles of evolutionary theory and the mechanisms of natural selection as applied to organismal populations. He describes new evolutionary ideas such as quantum speciation and discusses alternative phylogenies for relating species descent. Other topics include the role of chromosomal aberrations in evolution, the adaptive radiation of new species, the distribution of traits within a species (bird beak shapes, for example) along a cline, and the evolution of sex.

____________. *The New Evolutionary Timetable: Fossils, Genes, and the Origin of Species*. New York: Basic Books, 1981.
This book is a general survey of major topics and controversies in evolution. It discusses the Darwinian-Wallace Theory of Evolution and the challenges that this theory has faced, including creationism. It considers slow, gradualistic evolution as compared to rapid, punctuated evolution. Emphasis upon the fossil record data and molecular evolutionary data (such as DNA, RNA, and protein sequences) illustrates the methods of modern evolutionary science.

Stebbins, G. Ledyard. *The Basis of Progressive Evolution*. Chapel Hill: University of North Carolina Press, 1969.
Based upon a series of scholarly lectures, Stebbins briefly and eloquently summarizes the major pieces of evidence supporting our current scientific views on the origin and evolution of life on earth. He presents the elementary basics of genetics, the biochemical processes that occur within living cells, and cellular structures and cell physiology. Chapter 5, "The Evolution of Social Organization," makes some daring suggestions for cultural control, population restriction, and the modification of human society.

____________. *Processes of Organic Evolution*. Englewood Cliffs, N.J.: Prentice-Hall, 1966.

Stebbins' book is a general survey of evolution. It begins with a discussion of the development of the modern synthetic evolutionary theory. It describes genetics, mutations in genes (DNA), natural selection for or against mutant genes, and the variability of genes within populations of organisms of any given species. Stebbins describes the hybridization of species in plant evolution. He also specifically addresses the evolution of primates and humans.

Strickberger, Monroe W. *Evolution*. New York: Jones & Bartlett, 1990.

This book is a thorough examination of evolution, its history, methods of study, and mechanisms for its occurrence. The book is lengthy and detailed in its coverage, although Strickberger covers the subject material with great clarity. Numerous examples and illustrations are provided, along with references for further study. Covered evolutionary mechanisms include mutation, genetic drift, prezygotic and postzygotic species isolating mechanisms, geographical separation, and population migrations. The book is an excellent reference for all aspects of evolution.

Tax, Sol, ed. *Evolution After Darwin*. Vols. 1-3. Chicago: University of Chicago Press, 1960.

This three-volume collection of essays and discussions marked the centennial of Charles Darwin's 1859 publication of *On the Origin of Species*. The work is assembled from the proceedings of a November, 1959, conference of forty top evolutionary scientists at the University of Chicago. Volume 1 discusses the origin of life; Volume 2 discusses the biological and cultural evolution of humankind; Volume 3 discusses important topics and controversies in evolution. Diverse scientific disciplines herein represented include biology, chemistry, astronomy, and psychology.

Todes, Daniel P. *Darwin Without Malthus: The Struggle for Existence in Russian Evolutionary Thought*. New York: Oxford University Press, 1989.

Todes historical study analyzes the development of evolutionary thought in czarist Russia during the nineteenth century. This book represents a scholarly analysis of the impact of Darwinian evolution upon a repressive society that was beginning to feel the murmurs of a coming class struggle. The ideas and works of several Russian biologists, some of them political radicals, are described, people such as Andrei Nikolaevich Beketov, the "father of Russian botany," Sergei Ivanovich Korzhinskii, Il'ia Il'ich Mechnikov, K. F. Kessler, and Petr Alekseevich Kropotkin. The Darwinian struggle for existence as incorporated in leftist and communist class-struggle ideologies is emphasized.

Vermeij, Geerat J. *Evolution and Escalation: An Ecological History of Life*. Princeton, N.J.: Princeton University Press, 1987.
In this book, zoologist Vermeij describes the last six hundred million years of earth life, a period of time during which multicellular eukaryotes (such as protists, fungi, plants, and animals) evolved. He describes anatomical and physiological changes in species over time, from which he derives various patterns and trends in eukaryotic evolution. He proposes a theory of escalation in which evolution proceeds by changes in organisms and proliferation of these same organisms in response to ever-changing environmental conditions. He draws upon a wealth of information to support this view.

Volpe, E. Peter. *Understanding Evolution*. Dubuque, Iowa: Wm. C. Brown, 1967.
Volpe's evolution book is a short, concise introduction to the subject. All major evolutionary topics are discussed, including population biology, the genetics of populations, and the applications of the Castle-Hardy-Weinberg equilibrium equations to organismal populations. Volpe also describes genetic drift, the definition of species, varieties of species, and adaptive radiation of species into new environments. He presents the reader with several major evolutionary problems to illustrate his points, including sickle-cell anemia and the transmission of genes within populations. Numerous data tables and graphs are included in the text, along with bibliographies.

Wallace, Bruce. *Chromosomes, Giant Molecules, and Evolution*. New York: W. W. Norton, 1966.
Wallace, one of the premier evolutionary scientists of the molecular biology era, presents the various concepts and ideas of evolution in a very clear, concise fashion. He stresses mutations in genes and alterations in chromosome structure as the principal triggers in evolutionary change leading to the formation of new species. Each subject area is well illustrated with numerous examples and useful reference lists. The various mechanisms of natural selection, and how they influence genes and chromosomes, are discussed.

Chapter 2
EVIDENCE SUPPORTING EVOLUTION

Geology and Paleontology

Audley-Charles, M. G., and A. Hallam, eds. *Gondwana and Tethys*. New York: Oxford University Press, 1988.
This compilation of papers by leading evolutionary geologists looks back to about 250 million years ago, when all the land masses on earth were one single supercontinent that scientists call Pangaea. The two principle subdivisions of this supercontinent, Gondwana and Tethys, separated due to tectonic forces in the earth, thus triggering the oppositely directed drifts of the continents that we witness today. The papers deal with the structural and magnetic geology of Pangaea according to geological evidence. Some papers also consider the effects of the supercontinent breakup upon various life forms that became extinct at roughly the same time.

Bally, Albert W., and Allison R. Palmer, eds. *The Geology of North America*. Boulder, Colo.: Geological Society of America, 1989.
This lengthy book is a thorough discussion of the geological evolution of the North American continent. Twenty papers from noted geologists describe various factors and periods in the evolution of the continent, including seismic structure, sedimentation and rates of sedimentation, plate tectonic activities, and fossil fuel (such as coal or petroleum) distributions. The book contains informative illustrations and is accompanied by a boxed set of North American geological maps.

Berger, W. H., V. S. Smetacek, and G. Wefer, eds. *Productivity of the Ocean: Present and Past*. New York: Wiley-Interscience, 1989.
This compilation of research papers by forty-three leading paleoceanographers is aimed at studying the effects of climate upon ocean species during the earth's past geological history. Evidence indicates that the earth's atmosphere during the ice ages had a carbon dioxide content thirty times lower than today. This fact has serious implications with respect to global warming. The researchers focus upon the distribution of ocean species over geological time, most notably the microscopic phytoplankton that produces most of the world's food. The essays provide a detailed summary of available knowledge on climate and the evolution of the oceans.

Carroll, Robert L. *Vertebrate Paleontology and Evolution*. San Francisco: W. H. Freeman, 1988.
This enormous volume, compiled by a McGill University paleontologist, is an accurate, up-to-date collection of fossil evidence for the vertebrates, animals

having bony/cartilaginous endoskeletons. An informative text is supported by hundreds of sketches that accurately portray the anatomies of extinct and extant species. Carroll writes with authority and clarity. His book is smoothly organized using an evolutionary and class arrangement. The approximate periods of existence of various species also are chronicled.

Cloud, Preston. *Oasis in Space: Earth History from the Beginning*. New York: W. W. Norton, 1978.
Cloud's work is an exciting introduction to geology and paleontology for the beginner. He describes the geological history of earth and the evolution of life on earth during the past 4.6 billion years of the planet's history. In chapter 1, "The Nebula and Planets," Cloud describes the theories of solar system formation. In chapter 3, "On Reading the Rocks," he describes various geological techniques and experiments for analyzing field data. Cloud describes the evolution of life on earth, fossil data, extinctions of species, and appearances of new species in chapter 15, "Biological Change on a Mobile Earth: Evolution and Extinction."

Cope, Edward Drinker. *The Origin of the Fittest and the Primary Factors of Organic Evolution*. New York: Arno Press, 1974.
In this work, Cope considers a variety of evolutionary topics from both biological and metaphysical viewpoints. He discusses evolutionary theory and the origin of life, the definition of species, and the origins and classification of genera. He also examines the evolution of animal motion, mammalian teeth and feet, possible evolutionary phylogenies, the concept of "will," and kinetogenesis.

Dott, Robert H., Jr., and Roger L. Batten. *Evolution of the Earth*. Maps and diagrams by Randall D. Sole. New York: McGraw-Hill, 1981.
Dott and Batten's six-hundred-page introduction to the geological and evolutionary history of earth covers all geological eras, including the important periods during which life has flourished on the planet. Chapter 2, "Floods, Fossils, and Heresies," looks at objective geological experimentation as compared to the hearsay maintained by critics of modern geology. Chapter 5 describes the "Origin and Early Evolution of the Earth." Major geological changes on earth are described in chapter 6, "Mountain Building and Drifting Continents." The evolution of living organisms on earth and their fossil evidences are presented in chapter 9, "Early Life and Its Patterns." Good photographs and maps accompany the text.

Eldredge, Niles, and Joel Cracraft. *Phylogenetic Patterns and the Evolutionary Process*. New York: Columbia University Press, 1984.
In this excellent work, two leading biologists discuss a major point of contention in evolutionary theory: the construction of phylogenies that accurately portray the evolutionary relatedness of different species. The authors define species, describe the methods of biological classification of different organisms, and

describe the construction and analysis of cladograms/phylogenetic trees. Many examples, problems, and references are provided for further research.

Gregory, William K. *Evolution Emerging: A Survey of Changing Patterns from Primeval Life to Man*. New York: Macmillan, 1951.
This mammoth work is a textless collection of sketches, diagrams, and possible phylogenies illustrating the body plans and proposed evolution of invertebrate and vertebrate animal species. Fossils of extinct organisms and structures of existing organisms are compared and contrasted to suggest possible courses of evolution. This thousand-page book contains hundreds of illustrations.

__________. *Our Face from Fish to Man: A Portrait Gallery of Our Ancient Ancestors and Kinsfolk Together with a Concise History of Our Best Features*. New York: G. P. Putnam's Sons, 1929.
Despite the unusual title, this work is a compilation of paleontological and comparative anatomical evidences for evolution by a leading Columbia University paleontologist. Gregory looks at the evolutionary record, numerous fossil remains, the comparative anatomy of existing and extinct organisms, and the comparative embryological development of different species to sketch the progress of animal evolution from fish through humans.

Hoffman, Antoni. *Arguments on Evolution*. New York: Oxford University Press, 1989.
In this mentally stimulating volume, evolutionist Hoffman analyzes various concepts and approaches in paleontology and evolution such as the neutral theory of molecular evolution proposed by Motoo Kimura, species selection, and the periodicity of mass extinctions. In so doing, he questions the available data that supposedly supports certain views of evolutionary change. His philosophical approach forces the student to keep an open mind when judging the merits of competing evolutionary theories. Hoffman supports the neutral theory and maintains that even very small evolutionary changes must be considered in the process of speciation.

Joysey, K. A., and A. E. Friday, eds. *Problems of Phylogenetic Reconstruction*. New York: Academic Press, 1982.
This book is a collection of scientific papers presented at a Cambridge, England, meeting of the Systematics Association. Among the participating scientists were L. B. Halstead, R. A. Crowson, C. R. Hill, and C. Patterson. The participants discuss the problems of constructing phylogenies that accurately relate distinct species. The accuracy of the fossil record, the identification of key morphological characteristics of organisms, the uses of computers in constructing phylogenetic trees, and several possible species phylogenies are considered in the various papers.

Mason, Frances, ed. *Creation by Evolution*. New York: Macmillan, 1928.
This compilation of essays by leading early twentieth-century evolutionary scientists makes a very strong case for the Darwinian theory of evolution. Contributors to the volume include Julian S. Huxley, Henry F. Osborn, and William M. Wheeler. They discuss various sources of evidence supporting evolution, including paleontology (fossils) and comparative anatomy of species. They provide examples of evolutionary change for many different species, including humans, horses, birds, ants, moths, and plants.

Newman, Horatio Hackett. *Readings in Evolution, Genetics, and Eugenics*. Chicago: University of Chicago Press, 1921, 1925.
In this lengthy compendium of evolutionary evidence, Newman describes evolutionary theory and its mutational basis. He presents morphological evidence, comparative embryological evidence, and fossil evidence for the occurrence of evolution in a wide range of species. Numerous animal and plant species examples are described. Newman also discusses antievolution hysteria in the United States (such as the Scopes trial).

Nitecki, Matthew H., ed. *Evolutionary Progress*. Chicago: University of Chicago Press, 1988.
This work is a collection of essays and research papers by several major evolutionary scientists describing evidences for evolutionary change from numerous sources. Contributors to this volume include Francisco J. Ayala, John Maynard Smith, E. O. Wiley, David Raup, and Stephen Jay Gould. They discuss evidences from the fossil record, morphological changes in evolutionary history, physiological changes, complex species adaptations, and cultural evolution in humans.

Pecker, J. C., and S. K. Runcorn, eds. *The Earth's Climate and Variability of the Sun over Recent Millennia: Geophysical, Astronomical, and Archaeological Aspects*. London: The Royal Society, 1990.
This collection of research papers pools experts in astronomy and geology to illustrate the influence of the sun upon the earth's climate. The sun oscillates in activity over long periods of time. For example, our sun's sunspot cycle is approximately eleven years in rhythmic duration. Analysis of radiocarbon production in successively deeper geological strata has revealed a correlation between sunspot cycles and ice ages. Solar activity also may affect other earth processes, including carbon dioxide levels in the atmosphere, thereby indirectly influencing life on earth.

Ross, Robert M., and Warren D. Allmon. *Causes of Evolution: A Paleontological Perspective*. Chicago: University of Chicago Press, 1990.
Ross and Allmon look at the past five hundred million years of evolutionary history on earth principally from the fossil record. They examine the principal

changes that have occurred in both invertebrate and vertebrate animal species as well as changes in plant species. They consider important questions in evolutionary change such as whether species' genes or the environment are principally responsible for speciation, the appearance of new species on earth.

Schwarzbach, Martin. *Alfred Wegener: The Father of Continental Drift*. Madison, Wis.: Science Technology, 1986.
This extraordinary book describes the life of Alfred Wegener, the geologist and explorer who first proposed that the continents of earth are drifting, or floating, on the earth's interior upper mantle. Wegener's scientific work, research expeditions to Greenland and Iceland, and public reception of his theory are described with great clarity. Important discoveries (following his death) in paleomagnetism and geology that support his theory also are presented. Johannes Georgi and I. Bernard Cohen provide commentaries on Wegener's life and contributions.

Simpson, George Gaylord. *The Geography of Evolution: Selected Essays*. New York: Chilton Books, 1965.
Simpson's book is a collection of seven personal essays that had been published previously in scientific journals. These excellent summary papers describe evolution of several major groups of organisms and problems with the fossil interpretations of such evolution. In chapter 3, Simpson describes "Some Problems of Vertebrate Paleontology." He describes various aspects of mammalian evolution in chapters 5, "Historical Zoogeography of Australian Mammals," and 7, "Mammalian Evolution and the Southern Continents."

___________. *The Meaning of Evolution: A Study of the History of Life and of Its Significance for Man*. New Haven, Conn.: Yale University Press, 1949.
In this outstanding book, one of the foremost evolutionary scientists of this century describes the paleontological evidences for the occurrence of evolution. Simpson examines the geological time scale, the origins of life on earth, and the extensive wealth of fossils dating back to half a billion years. He also discusses the history of vertebrates (animals with bony or cartilaginous skeletons) with special emphasis on mammals. He also considers changes in evolutionary rates and factors that influence the speed and orientation of evolution.

Stearn, Colin W., Robert L. Carroll, and Thomas H. Clark. *Geological Evolution of North America*. 3d ed. New York: John Wiley & Sons, 1979.
Stearn, Carroll, and Clark's book is a valuable discussion of North American geology and fossilized species for the beginning student of the subject. They have included a great selection of diagrams and photographs. In chapter 1, "Continents and Oceans," they describe plate tectonics, continental drift, and the ancient supercontinent Pangaea. In chapter 3, "Stratigraphy," they discuss fossils and the interpretation of fossil finds. In chapter 5, "Time," they describe methods

for dating rock formations (such as radioactive carbon-14). They summarize major geological events in the evolution of life on earth period-by-period in chapter 8, "Life and Evolution."

Ziegler, Peter A. *Evolution of Laurussia: A Study in Late Paleozoic Plate Tectonics*. Norwell, Mass.: Kluwer, 1989.
In this publication for the Royal Geological and Mining Society of the Netherlands, geologist Ziegler describes the geological upheavals and plate tectonic events that occurred on earth from approximately 400 million years ago to 225 million years ago. During this time, the various continental plates drifted together to form a supercontinent that scientists call Pangaea. Europe and North America collided to form Laurussia, the beginnings of the supercontinent. The book is an exciting account of earth's history during this tumultuous time.

Ecology

Atchley, William R., and David S. Woodruff, eds. *Evolution and Speciation: Essays in Honour of M. J. D. White*. New York: Cambridge University Press, 1981.
This collection of scientific research papers by leading ecologists and evolutionary geneticists describes the species interactions and genetic evolutionary changes that have led to speciation, the development of new organismal species in nature. Example species include insects such as fruit flies (*Drosophila melanogaster*) and ticks as well as reptile and mammalian species. Several papers emphasize the role of a few major chromosomal rearrangements as driving forces in the appearance of new species. Contributing authors include James H. Oliver and Peter A. Parsons.

Bajema, Carl J., ed. *Artificial Selection and the Development of Evolutionary Theory*. Stroudsburg, Pa.: Dowden, Hutchinson & Ross, 1982.
This collection of essays, survey articles, and research papers spans the last few hundred years of evolutionary thought from a predominantly ecological perspective. Among the many illustrious authors are Francis Bacon, Charles Darwin, Sewall Wright, Joseph Priestly, Thomas R. Malthus, and Charles Lyell. Topics discussed include the theory of evolution, the mechanisms of natural selection and speciation, and the domestication of plants and animals. Numerous example species and processes are discussed along with relevant supporting data and references for further research.

Baker, H. G., and G. Ledyard Stebbins, eds. *The Genetics of Colonizing Species*. Proceedings of the First International Union of Biological Sciences Symposium on General Biology. New York: Academic Press, 1965.
This volume is a collection of research and survey papers that were presented

at a 1964 international ecology symposium in Asilomar, California. The list of contributing authors/scientists is impressive and includes Richard C. Lewontin, Ernst Mayr, C. H. Waddington, Edward O. Wilson, Frank Wilson, and Theodosius Dobzhansky. Among the many topics discussed are faunal dominance, self-pollinating plants, the origin of weeds, tropical crops, mammalian-marsupial interactions in Australia, genetic drift of natural populations, and the evolution of various species such as wheat and fruit flies (*Drosophila*).

Barbosa, Pedro, and Deborah K. Letourneau, eds. *Novel Aspects of Insect-Plant Interactions*. New York: Wiley-Interscience, 1988.
In this book, Barbosa and Letourneau describe the complex interactions between herbivorous insects and plants. Articles included in this volume examine both symbiotic and antagonistic relationships, the effects of plant chemicals upon insects, the manipulation of plant chemicals by insects, both Batesian and Mullerian mimicries, and the effects of plant chemicals upon carnivorous animals, high up the food chain, that have fed upon the insects. The evolutionary ramifications of insect-plant interactions are seen throughout the text.

Barigozzi, Claudio, ed. *Mechanisms of Speciation*. Proceedings of the International Meeting on Mechanisms of Speciation Sponsored by the Academia Nazionale dei Lincei, May 4-8, 1981, Rome, Italy. New York: Alan R. Liss, 1982.
This book is a collection of research papers and survey articles by leading evolutionary biologists and ecologists. Among the many contributed papers to this volume include Ernst Mayr's "Processes of Speciation in Animals," G. Ledyard Stebbins' "Plant Speciation," Francisco J. Ayala's "Gradualism Versus Punctualism in Speciation," and Alan R. Templeton's "Genetic Architecture of Speciation." Other papers discuss examples of evolution in unisexual fish, marine copepods, wheat, mammals, cave animals, and so forth.

Bendall, D. S., ed. *Evolution from Molecules to Men*. New York: Cambridge University Press, 1983.
All aspects and sources of evidence for evolution are presented in this collection of essays and research papers by leading authorities in ecology, genetics, paleontology, and physics. Sociobiology, the evolution of behavior, comparative DNA and protein sequencing between closely and distantly related species, the fossil record, and plate tectonics are among the many fascinating sources of evolutionary evidence that are examined in these articles. Scientists such as Ernst Mayr, Thomas Henry Huxley, Edward O. Wilson, Stephen Jay Gould, and Richard C. Lewontin have contributed to this volume.

Brower, Lincoln P., ed. *Mimicry and the Evolutionary Process*. Chicago: University of Chicago Press, 1988.
This collection of essays by leading evolutionary biologists considers the genetic,

behavioral, and ecological aspects of mimicry, a process by which some species evolve visual or behavioral cues that mirror those of other species. Mimicry helps species by protecting them from predators, by serving as a warning device, and even by assisting in the mating process in some unusual instances. The contributors to this volume examine the evolutionary mechanisms and adaptations of species mimicry.

Campbell, K. S. W., and M. F. Day, eds. *Rates of Evolution*. Boston: Allen & Unwin, 1987.
Papers presented at an international symposium on the mechanisms and speed of evolutionary change constitute this outstanding contribution to the literature. David M. Raup, M. R. Walters, Alan R. Templeton, and George L. G. Miklos head an impressive list of presenters. Evolutionary subject areas range from genetics to anatomy to geology. Special topics such as comparisons of evolutionary rates in prokaryotes (bacteria) versus eukaryotes (animals, plants, fungi, or protists), microevolutionary events versus macroevolutionary events, stasis, and punctuated equilibrium highlight this volume.

Christiansen, F. B., and T. M. Fenchel, eds. *Measuring Selection in Natural Populations*. Lecture Notes in Biomathematics 19. New York: Springer-Verlag, 1977.
Ecological and evolutionary genetics papers from a South Jutland symposium on population biology highlight this work, which is dedicated to the memory of Professor Ove Frydenberg. Many prominent evolutionary scientists have contributed to this volume, including Francisco Ayala, Luigi Cavalli-Sforza, John H. Gillespie, and Michael T. Clegg. Subject areas presented during the symposium include the evolution of sexual reproduction in eukaryotes, the distribution of various enzymes in natural populations of organisms, polymorphisms (traits influenced by many different genes), protein evolution in fruit flies, bird territoriality, coevolution of species (such as predator-prey and host-parasite), and human evolution. Several articles are very mathematical in concept and coverage.

Creed, Robert, ed. *Ecological Genetics and Evolution: Essays in Honour of E. B. Ford*. New York: Appleton-Century-Crofts, 1971.
Creed's book is a collection of research papers in ecology and evolution by leading scientists in the field, including Theodosius Dobzhansky, C. D. Darlington, and H. B. D. Kettlewell. Considerable supporting data is presented for evolution in a variety of species and processes. Major presentation topics include industrial melanism in the peppered moth *Biston betularia*, the occurrence of mimicry in a variety of species, especially butterflies, the occurrences of polymorphisms in many species, evolutionary evidence from the fossil record, and the evolution of plant species and insects.

Drake, J. A., et al., eds. *Biological Invasions: A Global Perspective*. New York: John Wiley & Sons, 1989.

This volume, published for the International Council of Scientific Unions Scientific Committee on Problems of the Environment, is a collection of papers by leading population biologists and ecologists. The theme of the book is the transfer of species from one location to another, thereby destabilizing the recipient environment. The invading species could be a parasite, predator, or serious competitor with indigenous species. Invasions here are studied for their role in evolutionary change; they also are studied for their impact when humans accidentally introduce certain species elsewhere. Numerous example situations are cited.

Endler, John A. *Geographical Variation, Speciation, and Clines*. Princeton, N.J.: Princeton University Press, 1977.

Endler's analysis of the formation of new organismal species is a simple, mathematical treatment of the subject. He describes the mechanisms of speciation (such as geographical isolation, genetic drift, gene flow, and mutation) within the context of the classic Castle-Hardy-Weinberg theory and equations for natural populations. Both laboratory and natural species populations are studied such as fruit flies and African passerine birds. Endler cites a host of experimental studies that constitute a lengthy reference list.

____________. *Natural Selection in the Wild*. Princeton, N.J.: Princeton University Press, 1986.

Another discussion of population biology describes gene pools within groups of organisms of any given species, natural selection acting upon the population gene pool, and mechanisms by which the population may split and eventually yield two or more distinct species. Mathematical modeling of population structures, such as the Castle-Hardy-Weinberg equations, and data tables are found throughout the text. The book contains interesting philosophical comments, good discussions of key points, and numerous experimental examples. The three modes of natural selection (directional, stabilizing, and disruptive) are clearly explained. Endler includes a special section covering methods of identifying the occurrence of natural selection in a population, something that is very difficult for the experimenter to accomplish.

Feldman, Marcus W., ed. *Mathematical Evolutionary Theory*. Princeton, N.J.: Princeton University Press, 1989.

This book is a collection of survey articles and actual research papers by a number of prominent mathematical ecologists, including John H. Gillespie, Bruce Weir, and Norman L. Kaplan. The papers are contributed in honor of the mathematical ecologist Samuel Karlin. They vary in degree of mathematical difficulty, although the descriptions of major evolutionary and ecological concepts are quite clear. Principal topics include birth, death, and migration, industrial melanism, marine organismal life cycles, repetitive DNA, and lactose absorption in mammals.

Forey, P. L., ed. *Chance, Change, and Challenge: The Evolving Biosphere.* New York: Cambridge University Press, 1981.
In association with the British Museum, Forey has assembled a collection of scientific articles covering diverse aspects of evolutionary ecology written by leading population biologists such as P. M. Hammond and R. L. Blackman. These clear, concise essays cover specialized topics such as mechanisms of speciation (including mutation, migration, genetic drift, and geographical isolation), distributions of different morphs (forms or races) of a species along geographical clines, polyploidy (multiple copies of chromosomes), and chromosome abnormalities. A variety of species examples are presented to illustrate the frequency of species-isolating mechanisms.

Futuyma, Douglas J., and Montgomery Slatkin, eds. *Coevolution.* Sunderland, Mass.: Sinauer Associates, 1983.
This work is another collection of evolutionary papers by scientists from a wide variety of disciplines ranging from ecology to paleontology to genetics. Contributing authors include Futuyma, Geerat J. Vermeij, Robert T. Bakker, and Lawrence Gilbert. Major paper topics include species mimicry, the evolution of helminths (parasitic worms) and their hosts, the coevolution of plants with their insect parasites-predators, the evolution of plants and fungi, and the interactions of bacteria, viruses, and plasmids with their various hosts. The emphasis of these papers is the simultaneous interactions and evolution of predator-prey and host-parasite species pairs.

Gause, G. F. *The Struggle for Existence.* New York: Hafner Press, 1964.
In this book, a leading Soviet zoologist discusses the mathematical aspects of evolution in a relatively simple fashion. He uses the important Lotka-Volterra equations for illustrating ecological relationships and coevolution between different species. He describes predator-prey species interactions, host-parasite species interactions, competition, mutualism, and so forth for a variety of organisms, including insects and plants as well as yeast and amoebae.

Gilpin, Michael E. *Group Selection in Predator-Prey Communities.* Princeton, N.J.: Princeton University Press, 1975.
Gilpin describes examples of predator-prey species interactions and the evolution of such interspecies systems in this brief but effective presentation. He discusses the theory behind such situations throughout nature, and he details mathematical models for describing such systems over time, even to the point of accurately predicting future interactions. His models of the effects of predator species upon prey species, and vice versa, enable the performance of simulations in the

laboratory and on computer. These models of coevolution can be accurately applied also to real-world species interactions.

Grant, B. Rosemary, and Peter R. Grant. *Evolutionary Dynamics of a Natural Population: The Large Cactus Finch of the Galapagos*. Chicago: University of Chicago Press, 1990.
In this book, the Grants describe their study of the cactus finch, one of fourteen different finch species first studied on the Galapagos Islands by Charles Darwin. They studied cactus finches on the Galapagos Island of Genovesa. From their studies, they conclude that there is considerable variability in beak size and shape among individuals of this species. As a result, climatic changes can drastically affect the availability of food and thereby favor birds having particular beak shapes, with the genes encoding these shapes being passed on to future offspring.

Grant, Peter R. *Ecology and Evolution of Darwin's Finches*. Princeton, N.J.: Princeton University Press, 1986.
Grant begins from where Darwin finished with studies of the finches of the Galapagos Islands. There are fourteen distinct species of finch on these eastern Pacific Islands, each species looking almost identical to the others yet being adapted for different feeding habits. Grant describes the morphology of each species, their respective breeding behaviors, and their probable evolutionary patterns of descent. He also discusses how the Galapagos finches can be used as a model for studying evolution in other species.

Grant, Verne. *Organismic Evolution*. San Francisco: W. H. Freeman, 1977.
In this well-organized, easy-to-read book, Grant presents a very clear picture of the evolutionary process and its various mechanisms. He discusses many different topics ranging primarily from ecology to geclogy to genetics. He describes mutation, gene flow, species isolating mechanisms, the structure of populations and gene pools, geological time, evolutionary rates, the social evolution of primates, and evolutionary controversies.

____________. *Plant Speciation*. New York: Columbia University Press, 1971, 1981.
In this extensive and well-researched work, a leading evolutionary scientist discusses many ecological and genetic factors that are believed to contribute to the evolution of living organisms on earth. These factors include migration, isolation, reproductive patterns, hybridization, polyploidization, chromosome translocations, and gene mutation/rearrangements. Grant, as usual, provides a wealth of example situations, many species, and a lengthy reference list for further study.

Ho, Mae-Wan, and Sidney W. Fox, eds. *Evolutionary Processes and Metaphors*. New York: John Wiley & Sons, 1988.
This volume pools the thoughts of ecologists, philosophers, psychologists, and mathematicians on the processes of evolution. Among the contributors are Walter M. Fitch and Kim Upper, who discuss the evolution of life and the genetic code, C. A. Cullis, who describes higher plant variation, B. C. Goodwin, who considers morphogenesis and heredity, R. Craw and R. Page, who examine panbiogeography, and Peter T. Saunders, who debates the merits of sociobiology theory. This outstanding book is well referenced.

Jackson, Jeremy B. C., Leo W. Buss, and Robert L. Cook, eds. *Population Biology and Evolution of Clonal Organisms*. New Haven, Conn.: Yale University Press, 1985.
This book is a collection of papers by several population biologists that concentrates upon the role of clonal growth as a principal evolutionary process. Since the origin of eukaryotic organisms more than one billion years ago, clonal growth has been a major feature of life on earth. Clonal growth is the basis of all higher organisms, including all animals and plants. Articles in this volume look at several species examples, including land plant (such as *Tilia*) and marine invertebrate animals (such as coral). The book includes a strong glossary and a good reference list.

King, James C. *The Biology of Race*. Berkeley: University of California Press, 1981.
In this informative work on a rather nebulous aspect of evolutionary theory, King explores major variations within different species that give rise to distinct populations having different physical traits. Such populations are called races, or morphs. They represent a small degree of evolution within each species, although not enough of a change to make the different populations into different species by preventing interbreeding and the production of fertile offspring. King looks at different races within species of birds, insects, and humans. He examines polygenic models for racial variations within these species. For humans, he looks at enzyme differences on the molecular level as well as differences in behavior, language, and social structure.

Krebs, Charles J. *Ecology: The Experimental Analysis of Distribution and Abundance*. New York: Harper & Row, 1978.
This book is one of the best in its field; it is an invaluable reference tool for ecologists, evolutionary scientists, and even laypeople interested in the environment. It provides an introduction to ecology and then describes the methods of ecological analysis. Hundreds of experimental studies are cited and described, particularly cases involving species interactions and the influences of physical/chemical factors upon the adaptiveness/survival of populations and entire species within specified areas. Numerous references are listed.

Krebs, J. R., and N. B. Davies. *An Introduction to Behavioural Ecology.* Sunderland, Mass.: Sinauer Associates, 1981.
In this book, Krebs and Davies examine the role of animal behavior in species interactions and evolution. They present relevant data from a variety of species, including birds, mammals, fish, and insects. They also present diagrams and mathematical models to illustrate species interactions (predator-prey and competition). Primate social organization is discussed in chapter 2, "Ecology and Adaptation: Comparisons between Species," and in chapter 7, "Parental Care and Mating Systems." Chapter 12, "Co-Evolution and Arms Races," examines predator-prey species interactions.

Lack, David. *Ecological Isolation in Birds.* Cambridge, Mass.: Harvard University Press, 1971.
Lack examines the worldwide distributions of various species of birds. In so doing, he concentrates upon groups of birds that are believed to be closely related in evolutionary time. By looking at the geographical distributions of closely related bird species, he attempts to arrive at possible models of migration and isolation that led to the evolution of these different species from one or a few common ancestors. Among the species surveyed include passerines, finches, nuthatches, hawks, vultures, sea gulls, Galapagos finches (fourteen species), sea gulls, and chickadees (twelve species). Special attention is given to island and continental isolating mechanisms of bird species formation. Each bird species is beautifully sketched for identification.

Manly, Bryan F. J. *The Statistics of Natural Selection on Animal Populations.* New York: Chapman & Hall, 1985.
This excellent source of ecological experimentation and mathematical modeling in evolutionary studies provides evidence for the occurrence of natural selection in populations from 1910 through 1959. Manly describes the important mark-recapture experiments, a technique in which a large number of individuals of a given species (such as pigeons) are captured, tagged, and released; later, individuals are recaptured and counted, both tagged and untagged, to obtain an estimate of various changes in the population's growth (such as birth rates, death rates, immigration, or emigration). Manly describes numerous case studies, analyzes gene frequency changes, and models populations using the Castle-Hardy-Weinberg equilibrium equations and other mathematical methods. Numerous charts, graphs, and references are provided in this thorough study.

Mayr, Ernst, and William B. Provine, eds. *The Evolutionary Synthesis: Perspectives on the Unification of Biology.* Cambridge, Mass.: Harvard University Press, 1980.
This collection of botanical, zoological, and paleontological essays by leading evolutionary scientists such as Theodosius Dobzhansky and C. D. Darlington

combines sources of evidence for evolution from different disciplines. The development of evolutionary theory is described in terms of mounting evidence from these fields and despite opposition from various misinformed groups. Mayr and Provine provide an interesting perspective on the development of evolution from scientists in different countries (such as the United States, Soviet Union, Great Britain, and France).

Mettler, Laurence E., and Thomas G. Gregg. *Population Genetics and Evolution*. Englewood Cliffs, N.J.: Prentice-Hall, 1969.
This theoretical and mathematical treatment of evolutionary theory and mechanisms provides clear explanations of major evolutionary concepts such as chromosome abnormalities (including inversions, deletions, and translocations) that contribute to speciation, the geomorphological isolation of populations leading to new species, and genetic variation and drift of traits within species populations. Mettler and Gregg cite many good examples of evolution in natural populations, including genetic variations within Darwin's Galapagos finches. The text is supported by numerous graphs, charts, and reference maps.

Murray, James. *Genetic Diversity and Natural Selection*. New York: Hafner Press, 1972.
In this excellent population biology introduction, a University of Virginia biologist describes the mechanisms by which populations of organisms develop into new species. Murray describes variations within the frequencies of different genes within given populations of organisms of a species, polymorphisms, and the effects of various natural selection forces (genetic drift and gene flow) upon these variations. He also looks at inter- and intra-species interactions as evolutionary driving forces. Chapter 4, "The Evolution of Dominance," discusses the importance of this phenomenon in determining the adaptability/survivability of specific traits. Many diagrams and reference citations support the text. Murray also presents simple mathematical models of evolutionary change.

Nitecki, Matthew H., ed. *Coevolution*. Chicago: University of Chicago Press, 1983.
This compilation is a collection of research and theoretical papers on the ecological aspects of evolutionary theory. Several types of species interactions are examined, including host-parasite relationships, predator-prey interactions, and mutualism. Debates over the coadaptability of interacting species versus coevolution between these species are considered. Numerous animal and plant species examples are cited and backed up by an impressive reference list. Contributors to this collection of papers include Leigh M. van Valen, Simon A. Levin, Peter Glynn, and Everett C. Olson.

Otte, Daniel, and John A. Endler, eds. *Speciation and Its Consequences*. Sunderland, Mass.: Sinauer Associates, 1989.

This volume is a compilation of papers that were presented by approximately forty leading evolutionary biologists, many of them population biologists, at an evolution symposium held in Philadelphia. The scientists argue over the theoretical interpretation of certain evolutionary trends, the definition of species, and major driving forces in evolution. There are, however, informative papers on the formation of new plant species from hybridization and the influences of biogeography and ecology on new species formation.

Parsons, Peter A. *The Evolutionary Biology of Colonizing Species*. New York: Cambridge University Press, 1983.
In this important view of evolution and species formation, Parsons describes the ecological succession of new species moving into previously unoccupied habitats. Such colonizing situations involve gene pools, natural selection, the evolution of specialized characteristics, competition between species, regulation of population sizes by available resources, predator-prey interactions, parasite-host interactions, and so forth. Parsons illustrates all of these concepts and others with great clarity and numerous example situations from the experimental research literature. He also describes species specializations and domestication of animals and plants.

Patrick, Ruth, ed. *Diversity*. Benchmark Papers in Ecology 13. Stroudsburg, Pa.: Dowden, Hutchinson & Ross, 1983.
Reprints of important ecological and evolutionary research papers are presented which describe evolutionary mechanisms and evidence for evolution from species such as ants, lizards, birds, bacteria, and protozoa. R. H. MacArthur describes "Patterns of Species Diversity." R. Patrick discusses the evolution and interactions of diverse species in "Stream Communities." The many articles contain excellent diagrams and are well referenced.

Patterson, Colin. *Evolution*. Ithaca, N.Y.: Cornell University Press, 1978.
This book, written by a leading evolutionary scientist under the auspices of the British Museum, provides a bountiful supply of evidence supporting evolution from many diverse scientific disciplines such as paleontology, ecology, and comparative anatomy/biochemistry. Patterson describes the DNA molecule, mutation, heredity, natural selection mechanisms, and the evolution of eukaryotic organisms (such as animals, plants, fungi, and protists). He uses several ecological models such as the classic examples first studied by Charles Darwin on the Galapagos Islands (the Galapagos finches and tortoises, for example). He also discusses the controversial sociobiology theory of Edward O. Wilson, in which the social evolution of mammals and insect societies, among others, is an extension of these species' genetic makeup and evolution.

Plotkin, H. C., ed. *The Role of Behavior in Evolution*. Cambridge, Mass.: MIT Press, 1988.

This collection of theoretical papers on ecology and evolution assembles the ideas of biologists, anthropologists, and philosophers. Topics include aspects of animal behavior such as learning, niche construction, and the effects of environmental stimuli upon behavior. The articles include good schematic diagrams, summary flowcharts to model key ideas and theories, and references for further study. Contributing scientists to this volume include Plotkin, David Hull, R. I. M. Dunbar, Robert N. Brandon, and F. J. Odling-Smee.

Rosenberg, Norman J. *Microclimate: The Biological Environment*. New York: John Wiley & Sons, 1974.
This analytical book is a description of factors that can influence the survival and, therefore, the evolution of new species in the environment. Various experimental techniques and equipment are described for the study of the distribution of species and for the quantitative measurement of the physical and chemical factors affecting species distributions. Appropriate photographs, diagrams, and references support the text.

Roubik, David W. *Ecology and Natural History of Tropical Bees*. New York: Cambridge University Press, 1989.
Roubik's work is a detailed analysis of the ecology of tropical bee species. He describes each bee species' community structure and the behaviors of individuals and small groups within the bee community population. He also describes the effects of predation, parasitism, and disease upon each bee species. Evolutionary models for bee speciation and evolution are also examined.

Scudder, Geoffrey G. E., and James L. Reveal. *Evolution Today*. Proceedings of the Second International Congress of Systematic and Evolutionary Biology. Pittsburgh: Carnegie-Mellon University Press, 1981.
These presentation papers at a major conference on evolution incorporate the ideas of several major evolutionary ecologists and geneticists. Research papers by John Maynard Smith, D. H. Frankel, Charles G. Sibley, James E. Darnell, and others cover such diverse topics as comparative DNA sequencing between different species, allozyme variations within natural populations of insect species, arctic biota, and colonizing mechanisms of species with specialized adaptations. The papers are relatively easy to follow and contain good references for further study.

Shvarts, S. S. *The Evolutionary Ecology of Animals: Ecological Mechanisms of the Evolutionary Process*. Translated by Ayesha E. Gill. New York: Consultants Bureau, 1977.
Shvarts, who died in 1976, was a leading Soviet evolutionary biologist at the Urals Branch of the Academy of Sciences of the Soviet Union. This work, published for the first time in the West, is a leading Soviet scientist's perspective upon evolutionary theory and mechanisms. Shvarts describes the major character-

istics of each of the principal terrestrial vertebrate animal classes. He then proceeds to describe the population genetics of animal species populations and the probable mechanisms by which new animal species have arisen during the past few hundred million years. He utilizes experimental data from population age structures and spatial distributions to support his arguments. His reference citations are an excellent guide to the Soviet evolutionary literature.

Smith, John Maynard, ed. *Evolution Now: A Century After Darwin*. San Francisco: W. H. Freeman, 1982.
This collection of scientific papers by leading evolutionists all appeared in the prestigious scientific journal *Nature*. They are published herein in recognition of the 123 years of scientific research following the publication of Darwin's *On the Origin of Species*. These papers discuss the theories concerning the origin of life on earth, the theory of evolution, mechanisms of natural selection, speciation, and evidences supporting evolution from scientific disciplines such as paleontology, ecology, comparative anatomy and physiology, comparative DNA and protein sequencing between various species, and biogeography.

Stebbins, G. Ledyard. *Flowering Plants: Evolution Above the Species Level*. Cambridge, Mass.: Harvard University Press, 1974.
In this impressive survey by a leading evolutionary biologist, the evolution of plant species over the last one billion years is described. Stebbins examines the special adaptations that have made plants so successful on earth, factors such as cross-pollination and the development of seeds. He also looks at plant specializations, the adaptive radiation of plant species, the evolution of gymnosperms, the evolution of angiosperms (flowering plants that bear fruit), and the division of the angiosperms into monocots and dicots with reference to seed leaves. Stebbins supports his arguments with numerous examples and references.

Stone, A. R., and D. L. Hawksworth, eds. *Coevolution and Systematics*. Oxford, England: Clarendon Press, 1986.
This special volume for the Systematics Association was Stone's last work before his untimely death. This book assembles papers from English biologists concerning the ecological coevolution of species and their interactions. Leading topics include patterns of coevolution, plant parasites, aphid-plant associations, lice-mammal interactions, and the evolution of helminths (parasitic worms). Contributors include John A. Barrett, John N. Thompson, C. H. C. Lyal, and J. M. Cox.

Wallace, Alfred Russel. *Contributions to the Theory of Natural Selection: A Series of Essays*. New York: Macmillan, 1870.
This book, first published eleven years after Darwin's *On the Origin of Species*, is Wallace's account of the theory of evolution, for which he and Darwin share credit based upon their research papers. Like Darwin, Wallace presents evolution

strongly from an ecological perspective. Chapter 3, "Mimicry, and Other Protective Resemblances Among Animals," describes the coevolution of certain traits between species, most notably the swallowtailed butterflies of Malaya. Chapter 5, "On Instinct in Man and Animals," speculates on human evolution, animal evolution, and various animal adaptations such as bird nesting. Wallace emphasizes the occurrence of natural selection throughout his text.

Weller, J. Marvin. *The Course of Evolution*. New York: McGraw-Hill, 1969.
In this lengthy book, Weller clearly describes the theory of evolution, the mechanisms of speciation, and speculations on the origin of life on earth approximately 3.8 billion years ago. He provides considerable evidence supporting evolution, including ecological, paleontological, and biochemical sources. He describes the evolution of plants, marine invertebrates, vertebrate animals (such as fish, amphibians, reptiles, birds, and mammals), primates, and humans. The book contains beautiful sketches and diagrams.

White, Michael J. D. *Modes of Speciation*. San Francisco: W. H. Freeman, 1978.
This treatise on ecological and genetic aspects of evolution is very thorough and incorporates many good experimental examples. Chapter 1, "Species and Speciation," studies the events and environmental pressures that divide populations and contribute to the diversification of new species. Chapter 2 looks at species differences. Later chapters tackle the details of evolutionary change such as chromosomal differences, polyploidy (many chromosomes), asexual versus sexual reproduction, and sympatric and allopatric modes of speciation. The book is very thorough in its coverage of complex topics.

Wiens, John A. *The Ecology of Bird Communities*. New York: Cambridge University Press, 1990.
In this comprehensive work written by an authority in the field of avian (bird) ecology, the ecological distribution and evolution of bird species are studied. Wiens compiles an impressive amount of ecological data to identify major patterns in the origin and evolution of new bird species over geological time. While concentrating upon interspecies competition, Wiens concludes that numerous factors may interplay in the evolutionary history of birds.

Williams, George C. *Sex and Evolution*. Princeton, N.J.: Princeton University Press, 1975.
In this interesting theoretical work, Williams discusses the events that led to the evolution of sexual reproduction, one of the distinguishing characteristics of most eukaryotic organisms. He speculates upon advantages that sexual reproduction may have as compared to asexual reproduction, using many different species examples. He compares populations having high fertility versus those having low

fertility. Among the model species that he uses in describing the evolution of sex are aphids, rotifers, strawberry plants, corals, elms, and oysters.

Wilson, David Sloan. *The Natural Selection of Populations and Communities*. Menlo Park, Calif.: Benjamin/Cummings, 1980.
This innovative work, published in association with the Institute of Ecology at the University of California, Davis, interprets the abundance of life on earth as a single superorganism, a popular notion among many modern scientists. With excellent descriptions and simple mathematical models, Wilson simulates the effects of natural selection upon natural populations of organisms. He talks about important concepts such as group selection, structured demes, fitness, altruism, exploitation, interference, and individual welfare versus community welfare in organismal populations. His approach is refreshing, and he poses many important questions concerning the ecological evolution of populations and species.

Wright, Sewall. *Evolution: Selected Papers*. Edited by William B. Provine. Chicago: University of Chicago Press, 1986.
The reprinted writings and research experiments of one of the great twentieth-century evolutionary biologists, Sewall Wright, are presented in this compendium. Wright's major papers from 1922 through 1982 describe domestication and selective breeding in cattle populations, inbreeding, dominance relationships among selected physical traits in various species, variations in gene frequencies from population to population for a given species, and self-sterility mutant alleles of specific genes. The articles are easy to follow, although many of them do employ the necessary rigor of mathematical modeling and statistical analysis of data.

Wynne-Edwards, V. C. *Evolution Through Group Selection*. Boston: Blackwell Scientific Publications, 1986.
This important work, written by a leading evolutionary scientist, is an accurate, authoritative summary of major ecological events on earth that contribute to evolution and speciation. Chapter 4, "The Three Trophic Industries," is a major centerpiece to the book that describes the flow of energy/matter through the food webs of life from primary producers (such as plants or plankton) to vivivores (such as animals or protozoa) to decomposers (such as fungi or bacteria). He emphasizes the limitations to available food and resources for any population of organisms. Using primarily vertebrate animal examples, he describes consumer relationships (such as predator-prey), territory sizes, and group selection in structured populations. Excellent schematic diagrams support the text. A valuable reference work.

Yablokov, A. V. *Phenetics: Evolution, Population, Trait*. Translated by Marie Jaroszeuska Hall. New York: Columbia University Press, 1986.

Yablokov's alternate view to evolutionary mechanisms in natural populations concentrates upon "phenes" as the primary units of evolutionary selection. He defines a "phene" as a "trait marker of genetic composition." Essentially, Yablokov is maintaining that the objects of natural selection are specific organismal characteristics, not merely the genes that confer these traits. He cites evidence from many experimental sources, including Sewall Wright and Theodosius Dobzhansky, to build a theory of population biology that incorporates phene pools, phenogeography, and so forth. His work is an interesting approach to population genetics and ecology.

Comparative Anatomy and Physiology

Arthur, Wallace. *Mechanisms of Morphological Evolution: A Combined Genetic, Developmental, and Ecological Approach*. New York: John Wiley & Sons, 1984.
Arthur's book is an excellent introduction to evolutionary theory from both biochemical and ecological perspectives. He presents neo-Darwinian evolutionary theory with a discussion of mutation and natural selection for or against mutations. He looks at models of populations over time and the construction of phylogenies showing the relatedness of different species. Hominid evolution, variation in bird beak sizes, and artificial selection in mice and fruit flies are among the many other topics discussed. An extensive reference list supports the text.

Ayala, Francisco J., ed. *Molecular Evolution*. Sunderland, Mass.: Sinauer Associates, 1976.
This book is an authoritative collection of research papers and essays by leading molecular evolutionary biologists. Impressive DNA and protein studies provide compelling support for the synthetic Darwinian theory of evolution. Leslie D. Gottlieb describes plant speciation at the molecular level, Morris Goodman presents "Protein Sequences in Phylogeny," and Allan C. Wilson speculates on "Gene Regulation in Evolution." Other contributors to this valuable reference include Theodosius Dobzhansky, John C. Avise, James W. Valentine, and Michael Soule.

Barbieri, Marcello. *The Semantic Theory of Evolution*. New York: Harwood, 1985.
Barbieri describes the origin and evolution of life on earth strongly from the molecular level, with major events occurring within DNA and protein. Chapter 7, "The Ribotype Theory of the Origins," looks at the evolution of life first as living molecules that eventually developed into the chromosomes, enzymes, and complex ribosomes that we see in the complex machinery of modern cells. Barbieri emphasizes the DNA-to-RNA-to-protein central dogma of molecular biology as a central theme in his arguments. Barbieri also examines the "grammatical laws" of evolution and the controversial sociobiology theory of evolution.

Bonner, John Tyler, ed. *Evolution and Development*. New York: Springer-Verlag, 1982.

In this work, Bonner has assembled research and speculative papers on the molecular evolution of life from several leading scientists, including Eric H. Davidson, Thomas C. Kaufman, J. C. Gerhart, I. Dawid, and Stephen Jay Gould. The origin and evolution of DNA, genes, and the developmental control of genes during the development of living organisms are among the very complex problems tackled in these papers. Genome alternatives are examined along with the sizes and shapes of cells/organisms. Metazoan morphogenesis, the evolution of shape and form in animal body structure, is a principal theme throughout the papers and in the discussions, which were held at a special symposium in Germany.

Brandon, Robert N., and Richard M. Burian, eds. *Genes, Organisms, Populations: Controversies over the Units of Selection*. Cambridge, Mass.: MIT Press, 1984.

This interesting collection of philosophical and mechanistic papers on the molecular evolution of living organisms examines a wide range of issues, including mutation, inbreeding, cross-breeding, group selection of populations, kin selection, selection factors, and the social aptitudes of primates. Problems in these areas are debated, and research strategies are proposed for solving and testing evolutionary hypotheses. Contributors include Richard C. Lewontin, Sewall Wright, Ernst Mayr, and Richard Dawkins.

Cavalier-Smith, T., ed. *The Evolution of Genome Size*. New York: John Wiley & Sons, 1985.

This well-illustrated volume contains essays and research articles by leading evolutionary geneticists and biochemists who address some baffling molecular problems. Cavalier-Smith introduces the book with an exploration of the C-value paradox, a situation in which higher organisms have increasing amounts of DNA, although some amphibians have incredibly more DNA than they would seem to need; a few other species exhibit the same paradoxical excess of DNA. M. Herdman looks at the evolution of bacterial genes. G. D. Clark-Walker examines mitochondrial DNA, genes that once may have been part of a completely independent organism. N. Gillham explores the DNA of chloroplasts and other plastid organelles. The research articles contain illustrations, some mathematics, and numerous references.

Chai, Chen Kang. *Genetic Evolution*. Chicago: University of Chicago Press, 1976.

In this very clear and comprehensive survey of molecular biology and evolution, Chai extensively covers the molecular evidence that supports evolutionary theory. He begins with a discussion of DNA, proteins, mutation, polymorphisms, chromosome abnormalities, and genetic disease. Chapter 5, "Evolution of Structure and Function," describes the evolutionary development of higher eukaryotic organisms. He cites evidence from many sources, including blood-type data, antibody experiments, and comparative embryology. Many data tables, photographs, and reference lists are presented throughout the text.

Dawkins, Richard. *The Extended Phenotype: The Gene as the Unit of Selection*. San Francisco: W. H. Freeman, 1982.
As a sequel to Dawkins' *The Selfish Gene*, this book continues the author's view of life as occurring primarily at the genetic level. Dawkins views the organism as the means of perpetuating genetic material. As such, the cell, enzymes, and all aspects of the organism are geared to protecting and propagating DNA. Chapter 9, "Selfish DNA, Jumping Genes, and a Lamarckian Scare," reiterates these ideas with particular reference to the parasitic genes of viruses and prions, among others. Chapter 14, "Rediscovering the Organism," stresses the importance of our reassessing the nature and meaning of life at the molecular level.

Dawson, Peter S., and Charles E. King, eds. *Readings in Population Biology*. Englewood Cliffs, N.J.: Prentice-Hall, 1971.
This collection of classic scientific papers by many of the top evolutionary biologists of the twentieth century provides convincing evidence for the existence of evolution in many different sample species. The emphasis throughout these important works is on the genetics of populations. Detailed genetic analyses of evolution in such things as fruit flies, mice, barnacles, birds, and grass are presented along with several mathematical models for describing evolutionary change in these populations. An impressive list of contributors includes Theodosius Dobzhansky, Richard C. Lewontin, J. B. S. Haldane, and John Maynard Smith. Original journal citations appear with each scientific paper.

Dobzhansky, Theodosius, et al., eds. *Evolutionary Biology*. Vols. 1-25. New York: Plenum Press, 1967-1991.
This annually published series of five to ten review articles per volume is an extraordinary window to major evolutionary topics, problems, and current research efforts as well as a tremendous access tool to the evolutionary literature. Each published article is written by an authority in her/his field and exhaustively covers the subject matter. Lengthy reference lists end each article. Dobzhansky, Hecht, and Steere were the founding editors in 1967. Following Dobzhansky's death in 1975, Wallace became a coeditor. Prance replaced Steere as coeditor in 1981.

Dover, G. A., and R. B. Flavell, eds. *Genome Evolution*. New York: Academic Press, 1982.
This publication for the Systematics Association is a compilation of research and theoretical papers on the molecular evolution of living organisms. Many outstanding and critical issues are described in these papers, including "selfish" DNA, transposable DNA elements that move from chromosome to chromosome, the

mystery of human immunoglobulin genes and their control, DNA repeats, and conserved DNA sequences that are found within the cells of all eukaryotic species. The still up-to-date articles are by such scientists as Eric H. Davidson, Richard Flavell, John Maynard Smith, and George L. G. Mikles. Excellent diagrams accompany each paper.

Ehrlich, Paul R., Richard W. Holm, and Dennis R. Parnell. *The Process of Evolution.* New York: McGraw-Hill, 1963.
Despite its age, this book is a strong introduction to evolutionary theory at the molecular level. The authors, being leading biologists, were aware of the major breakthroughs in genetics that were occurring at the time of publication for this book. They describe the theoretical origins of life, the evolution of genes, and mutations within genes that led to the origin and evolution of living cells. They present a thorough account of the development and differentiation of tissues in plants and animals, events that are genetically controlled and that are complex products of evolution. Extensive examples and references are provided for further study.

Eigen, M., and P. Schuster. *The Hypercycle: A Principle of Natural Self-Organization.* New York: Springer-Verlag, 1979.
In this brilliant theoretical work, Eigen and Schuster describe the order and evolution of living organisms in terms of intricate molecular relationships and energy cycles through molecules. Their theory is a mathematically nonlinear explanation of evolution and of the arrangement of living organisms and of individual organisms. In terms of life, the authors describe several major hypercycles, including the critical and predominant DNA-to-RNA-to-protein sequence and the tricarboxylic acid cycle for energy transformation within cells. Many diagrams are provided to emphasize key points.

Feldman, Marcus W., ed. *Mathematical Evolutionary Theory.* Princeton, N.J.: Princeton University Press, 1989.
This complicated analysis of mathematical theory in biology is designed to stress the study of biological populations and of gene pools within populations. The compiled papers are written by several leading population biologists. Topics include the history of mathematical applications to evolutionary theory, competing mathematical theories, and controversies in evolutionary models.

Fox, J. Lawrence, Zdenek Deyl, and Anton Blazej, eds. *Protein Structure and Evolution.* New York: Marcel Dekker, 1976.
This book is a compilation of presentation papers and discussions for a June, 1975, symposium on molecular evolution in Czechoslovakia. Most of the participants are Eastern European scientists; this volume is a unique glimpse at evolutionary research behind the Iron Curtain prior to 1976. Paper topics include DNA

and protein sequence analysis, the genetic code, the origins of the first enzymes, hormones in gene control, immunology, and neurotoxins. The papers are well organized and clearly written.

Futuyma, Douglas J. *Evolutionary Biology*. Sunderland, Mass.: Sinauer Associates, 1979.
This book, written by a prominent SUNY-Stony Brook evolutionary biologist, is a clear introduction to evolution, especially at the molecular and enzymatic level. Futuyma describes evolutionary theory, evolutionary mechanisms, and genetic variation within natural populations. He looks at special topics such as the effects of natural selection upon gene frequencies, microevolution versus macroevolution, the measurement of evolutionary rates, and the construction of evolutionary phylogenies. The book is well illustrated with many examples and references.

Gershowitz, Henry, Donald L. Rucknagel, and Richard E. Tashian, eds. *Evolutionary Perspectives and the New Genetics*. New York: Alan R. Liss, 1986.
Presentation research papers at a University of Michigan symposium on molecular evolution highlight this volume honoring biologist James V. Neel. Coverage topics at the conference include globin gene mutations, the evolution of myoglobin and hemoglobin proteins, gene conversion mechanisms, comparative DNA and protein sequencing between various species, and constructions of possible molecular phylogenies relating different species on the evolutionary time scale. Contributing authors include Daniel Hartl and J. B. Clegg.

Goodman, Morris, ed. *Macromolecular Sequences in Systematic and Evolutionary Biology*. New York: Plenum Press, 1982.
This tremendous reference is a collection of lengthy, exhaustively researched summary papers in evolutionary theory. The articles, written by leading scientists in molecular evolutionary research, are thorough but extremely well written. Contributors include Goodman, Michael Coates, Richard Holmquist, Lois T. Hunt, Simon Stone, and Thomas H. Jukes. Covered research areas include higher eutherian phylogeny, primate amino acid sequence comparisons, vertebrate eye lens proteins, the evolution of chromosomal proteins, and mammalian pancreatic ribonuclease enzymes.

Grasse, Pierre P. *Evolution of Living Organisms: Evidence for a New Theory of Transformation*. New York: Academic Press, 1977.
Grasse, a University of Paris VI evolutionary scientist, proposes a type of evolution based upon punctualistic, or catastrophic, change, where inherent genetic components of organisms assist organisms in rapid evolutionary change. Grasse bases his proposals upon molecular evolutionary data. Chapter 2, "Creative Evolution, or the Appearance of Types of Organization," considers the

possibility of premammalian reptiles descending from cynodonts two hundred million years ago. Chapter 7, "Evolution and Necessity," reiterates his proposals that self-organizing groups of organisms capitalize upon rapid changes in their environments.

Gutfreund, H., ed. *Biochemical Evolution*. New York: Cambridge University Press, 1981.
This excellent survey of molecular evolutionary data includes research and theoretical papers from a number of prominent evolutionary biologists, including Peter Schuster and Alan Weeds. Molecular studies of immunoglobulins (antibodies) from humans and other species, muscle contractility molecules, visual pigments, photosynthetic molecules in plants, and bacterial enzymes provide strong arguments for the construction of various species' phylogenetic trees. Prebiotic evolution and the origin of life are also discussed. Excellent diagrams and reference lists accompany each paper.

Harbome, J. B., ed. *Biochemical Aspects of Plant and Animal Coevolution*. New York: Academic Press, 1978.
This collection of research papers from numerous plant molecular biologists examines the biochemical aspects of animal and plant interactions (such as mutualism, predator-prey, or parasite-host). The papers examine the chemicals produced by organisms involved in each type of interaction and how these chemicals affect other involved organisms. Topics include the fungal parasites of plants, plant biochemistry, seed toxins, and carotenoid photosynthetic pigments. Contributors to this volume include Miriam Rothschild and Gunnar Bergstrom.

Hillis, David M., and Craig Moritz. *Molecular Systematics*. Sunderland, Mass.: Sinauer Associates, 1990.
This reference work serves as an interface between molecular biology and the classification of organisms. The DNA nucleotide sequences of certain genes that are shared between different species can be compared to determine how closely related the different species may be in evolutionary time. The book describes a variety of techniques (such as gel electrophoresis, immunology, and DNA sequencing) for classifying organisms based upon DNA and amino acid sequence data.

Hurst, C. C. *The Mechanism of Creative Evolution*. New York: Macmillan, 1932.
In this very accurate and early look at the molecular evolution of living organisms, Hurst examines Darwinian evolutionary theory and links it to Mendel's laws of heredity. He also brings into consideration the mutational work of Thomas Hunt Morgan, whose mutational work with fruit flies clearly demonstrated that the genes located on chromosomes are the physical basis of heredity and evolution. Hurst describes the gene as the "unit of life." He describes various aspects

of evolutionary genetics, including hybridization, polyploidy, experimental mutations, chromosome maps, and the evolution of sex using many animal and plant species examples. The book is thoroughly researched and referenced.

Ishii, Susumu, Tetsuya Hirano, and Masura Wada, eds. *Hormones, Adaptation and Evolution*. New York: Springer-Verlag, 1980.
This volume is a collection of research papers presented mostly by Japanese evolutionary scientists at a 1979 international evolutionary biology symposium at the University of Tokyo. As such, it provides a glimpse of Japanese advances in the study of molecular evolution. Research papers herein cover the comparative anatomical and physiological analysis of various species ranging from mammals to birds to insects. Several detailed papers of interest examine the evolution of reproductive cycles and important rearrangements in cardiovascular physiology during vertebrate evolution.

Jameson, D. L., ed. *Genetics of Speciation*. Benchmark Papers in Genetics 9. Stroudsburg, Pa.: Dowden, Hutchinson & Ross, 1977.
Jameson has collected an impressive array of classic evolutionary genetics research papers in this outstanding volume. The papers are reprinted from scientific journals such as the *Journal of Genetics*, *American Naturalist*, *Evolution*, and *Proceedings of the National Academy of Sciences (USA)*. The many important articles include J. Schmidt's "Racial Studies in Fishes," Alan C. Wilson's pioneering work in molecular evolution, Francisco J. Ayala's studies on the genetics of speciation in the fruit fly *Drosophila melanogaster*, and J. B. S. Haldane's research into hybrid sterility. Other contributors include Alfred H. Sturtevant, H. L. Carson, Theodosius Dobzhansky, and Ernst Mayr.

Karlin, Samuel, and Eviatar Nevo, eds. *Evolutionary Processes and Theory*. Orlando, Fla.: Academic Press/Harcourt Brace Jovanovich, 1986.
This unique collection of very clear research papers came from a 1985 molecular evolution workshop in Israel. Papers by such eminent scientists as Bruce Levin, Michael T. Clegg, Alan R. Templeton, and Margaret G. Kidwell highlight this volume. Principal topics discussed at the workshop and in the papers include DNA and protein sequence comparisons between different species, chloroplast and mitochondrial DNA evolution, the evolution of behavior in mammals, and gene control mechanisms at the cellular level. The volume is a valuable reference for the student of evolution.

Leeper, G. W. *The Evolution of Living Organisms*. Melbourne, Australia: Melbourne University Press, 1962.
This collection of research papers from an international array of evolutionary scientists covers evolutionary theory from several different angles, including molecular evolution, comparative anatomy and physiology, and ecology. These

papers were presented at a 1959 symposium sponsored by the Royal Society of Victoria. Principal topics presented and discussed include the evolution of vertebrates from devonian fish to humans, thyroid gland evolution in vertebrates, chromosomal inversions in insects, and the evolution of species ranging from marsupials to crabs to plants.

Loeschcke, Volker, ed. *Genetic Constraints on Adaptive Evolution*. New York: Springer-Verlag, 1987.
This interesting collection of molecular and ecological evolutionary papers describes mutational changes in DNA, chromosomal aberrations, unusual DNA patterns and arrangements within the cells of various species, geographical separations of populations leading to speciation, pleiotropy (that is, one mutation affects many organismal characteristics), and phenotypic plasticity (the ability of a given organismal characteristic to respond to a variety of environmental/selective pressures). Contributing scientists include F. B. Christiansen, A. G. Clark, G. B. Golding, and A. J. van Noordwijk.

MacIntyre, Ross J., ed. *Molecular Evolutionary Genetics*. New York: Plenum Press, 1985.
This volume is a collection of up-to-date experiments in molecular evolutionary genetics. An extensive parade of research papers covers important concepts such as the origin and evolution of DNA sequences, repetitive DNA "families," highly repetitive DNA, ribosomal RNA-encoding genes, retroviruses, and the interesting evolution of mitochondrial DNAs from algae, plants, birds, reptiles, and mammals. Contributors include Wesley M. Brown, George L. G. Miklos, and Wen Hsiung Li.

Manwell, Clyde, and C. M. Ann Baker. *Molecular Biology and the Origin of Species*. Seattle: University of Washington Press, 1970.
By utilizing a large supply of experimental reference data, Manwell and Baker present an impressive quantity of molecular evidence supporting the occurrence of evolution in numerous species. They examine a number of concepts and experiments in the field, including gene mutation, gene duplication and divergent evolution, supergenes, heterosis, and the classification of proteins. Special attention is given to the evolution of bodily fluid proteins such as egg white protein (for example, albumin) and hemoglobin. The book contains extensive examples and references.

Matsuda, Ryuichi. *Animal Evolution in Changing Environments with Special Reference to Abnormal Metamorphosis*. New York: John Wiley & Sons, 1987.
This book is a phylum-by-phylum journey through the animal kingdom. In the process, Matsuda examines apparent abnormalities in the development of organisms of various species. He identifies abnormalities in both embryological

and adult development and differentiation. He considers genetic and/or environmental explanations for these defects, and demonstrates the flexibility/variability of genes to evolutionary changes brought about by changes in the environment (natural selection). Good illustrations and references support Matsuda's arguments.

Medvedev, Zhores A. *Molecular-Genetic Mechanisms of Development*. Translated by Basil Haugh. New York: Plenum Press, 1970.
This survey of molecular genetics concentrates upon the complex role of genes in the regulation of organismal development, from conception to full development. The complex developmental changes that occur in plants and especially in vertebrate animal species involve an intricately detailed system of perhaps thousands of genes that have evolved to work together with incredible precision. Medvedev describes the flow of information in the organismal cell from DNA to RNA to protein, the role of ribosomes in protein translation, gene control mechanisms at all stages of information transfer, and regulatory molecules (such as hormones). Chromosome evolution and alterations in DNA expression (that is, some genes turn on, others turn off) during cellular differentiation and morphogenesis are given special attention in this volume.

Morgan, Thomas Hunt, Alfred H. Sturtevant, H. J. Muller, and C. B. Bridges. *The Mechanism of Mendelian Heredity*. New York: Holt, Rinehart and Winston, 1923. Reprint. Johnson Reprint Corporation, 1972.
In this classic and still-relevant text, four giants of twentieth-century genetics combine their talents to describe heredity and the mechanisms of evolution. The concept of genes as the units of inheritance is explained along with solid evidence that they are located on cellular chromosomes. Chromosomes are discussed within the context of Mendelian inheritance. Genetic linkage on chromosomes, allelomorphs, chromosome abnormalities, experimentally induced mutations, and genetic maps of chromosomes for organisms such as the fruit fly *Drosophila melanogaster* are described along with their influences upon species evolution.

Nei, Masatoshi. *Molecular Evolutionary Genetics*. New York: Columbia University Press, 1987.
Nei's work is a five-hundred-page survey of the history of life on earth from a molecular viewpoint. Nei begins with the origin of DNA, the origin of living cells, the evolution of the genetic code found within the nucleotide sequence of DNA and the amino acid sequence of proteins, mutation and mutational effects upon DNA and encoded proteins, increases and changes in genome size, and the evolution of eukaryotic (many-celled) organisms. Nei explains the distribution and variation of genes within populations of organisms with simple mathematical models. From comparative DNA and protein sequencing experiments, phylogene-

tic trees are constructed to demonstrate probable relatedness between various species. The included data and references are up-to-date.

Nei, Masatoshi, and Richard K. Koehn, eds. *Evolution of Genes and Proteins.* Sunderland, Mass.: Sinauer Associates, 1983.
This informative collection of modern evolutionary papers, both experimental and theoretical in nature, describes major recent breakthroughs in molecular biology and evolution. The evolution of the genetic code, mutational rates of evolution, Motoo Kimura's neutral theory of evolution, transposable elements ("jumping" genes), antibody gene structure, globin gene structure, mitochondrial DNA, multigene families, and protein polymorphisms are among the many exciting research topics in this volume. Contributing scientists include Frank Burton, John C. Avise, John G. Hall, and Motoo Kimura.

Ninio, Jacques. *Molecular Approaches to Evolution.* Translated by Robert Lang. Princeton, N.J.: Princeton University Press, 1979, 1983.
This comprehensive survey of molecular evolution describes DNA, RNA, protein, the genetic code, and the evolution of molecules within the living cell that have contributed to the diversity of living forms on earth. The origins of life and of the genetic code are considered along with unique special molecules such as ribosomes and transfer RNA, the key components of the protein translational machinery in all living cells. Other topics include sequence comparisons of DNA molecules from different species, population genetics, the construction of phylogenetic trees to relate different species, and molecular crosses. Bioengineering and cloning are also discussed.

Nitecki, Matthew H., ed. *Biochemical Aspects of Evolutionary Biology.* Chicago: University of Chicago Press, 1982.
Nitecki's work is a collection of strictly biochemical research and survey articles that demonstrate evolutionary patterns within a variety of animal and plant species. Major covered topics include pheromones (hormones released between organisms for sex, territoriality, and so forth), the evolution of pollination mechanisms in plants, the photosynthetic evolution of drought-resistant C-4 plants, and specialized chemicals that have contributed to the evolution of plant and animal species over the last five hundred million years. Contributing scientists to this volume include Lynn Margulis, Karl J. Niklos, Stevan J. Arnold, and Francisco J. Ayala.

___________. *Evolutionary Innovation.* Chicago: University of Chicago Press, 1990.
Another collection of research articles on the evolutionary adaptations for survival that have been developed by various species incorporates the ideas of many leading scientists from diverse disciplines. The clearly written, well-illustrated

articles cover topics such as the comparative morphology (body forms) of both extant (living) and extinct species. The fossil record supplies considerable evolutionary evidence in several articles. Other articles concentrate upon the evolution of gene regulation in the development and differentiation of higher animal species. Very nice models of the evolutionary process are illustrated in a number of the articles.

Patterson, Colin, ed. *Molecules and Morphology in Evolution: Conflict or Compromise*. New York: Cambridge University Press, 1987.
A variety of relevant issues in molecular evolutionary theory are presented in this volume, which contains essays by leading geneticists and biochemists. The evolution of hominids (early humans) from primates, the accuracy of prokaryotic and eukaryotic evolutionary phylogenies, the evolution of avian (bird) DNA, the origins of tetrapods (four-legged animals), and microevolution versus macroevolutionary trends are outlined in numerous research papers. Contributors to this volume include C. R. Woese and Morris Goodman.

Raff, Rudolf A., and Elizabeth C. Raff, eds. *Development as an Evolutionary Process*. New York: Alan R. Liss, 1987.
This compendium of evolutionary papers contributed by leading biologists constitutes volume eight of the MBL Lectures in Biology. Several fascinating molecular evolutionary papers are presented on problems and ideas such as transposable elements (genes that move from chromosome to chromosome), multiple gene "families," and the molecular evolution of species such as fruit flies, frogs, brachiopods, nematodes, and plants. Contributors include H. Bruce Diaz, Fotis C. Kafatos, James W. Valentine, Victor Ambros, and William Fixsen.

Sang, James H. *Genetics and Development*. New York: Longman, 1984.
This invaluable, up-to-date reference work is a short survey of important genetic experiments aimed at understanding the development of individual organisms. At the same time, it provides us with glimpses of the evolutionary process in terms of the genes that are shared between many remotely related species. Chapter 1, "Development and Differentiation," describes organismal development and change over time, especially in animals. Chapter 5, "The Rate and Significance of Mutations," discusses the role of mutation in changing DNA and thereby driving evolution.

Sharma, A. K., and A. Sharma, eds. *Chromosomes in Evolution of Eukaryotic Groups*. Vol. 1. Boca Raton, Fla.: CRC Press, 1983.
This collection of five summary research articles from leading American evolutionary laboratories describes the recent major breakthroughs in molecular biology and evolution. Most of the articles deal with major chromosomal changes as driving forces in evolution, changes such as chromosomal inversions, transloca-

tions, and fusions. Gene variation and heterochromatin are also discussed. The examples, photographs, and reference lists are outstanding.

Simpson, George Gaylord. *The Major Features of Evolution*. New York: Columbia University Press, 1953.
In this lengthy volume, the eminent evolutionary biologist Simpson describes evolution at the molecular level, specifically from the viewpoint of the DNA molecule (the gene). Gene mutations, variations in gene frequencies within a population of organisms brought about by mutation, migration, and so forth, and natural selection for or against mutations are described with great clarity. Major trends in evolution, such as periodic mass extinctions and the evolution of bradytely and tachytely in animals, are considered as well.

____________. *Tempo and Mode in Evolution*. New York: Columbia University Press, 1944.
This introduction to evolutionary processes, written by one of the foremost twentieth- century authorities on the subject, describes evolution in a very simple fashion. Simpson describes the gene (DNA) as the unit of evolutionary change, variations in gene frequencies within populations of organisms, mutations in genes, and natural selection for or against variable genes based upon many factors (such as population size or migration). He also discusses aspects of micro-, macro-, and mega-evolution. He proposes a new quantum view of evolutionary change.

Smith, H. H., ed. *Evolution of Genetic Systems*. New York: Gordon & Breach, 1972.
Carl R. Woese, R. Sager, and Margaret N. Farquhar head the list of geneticists and biochemists who have contributed to this volume on evolutionary biochemistry. In several informative papers, they and their colleagues explore major issues in evolutionary genetics such as chromosomal rearrangements in evolution, the evolution of genome size and the genetic evidence for a reassessment of phylogenetic trees relating various bacterial species. The research articles are supported by extensive graphs and data tables.

Smith, John Maynard. *Evolution and the Theory of Games*. New York: Cambridge University Press, 1982.
In this interesting book, Smith, one of the leading evolutionary geneticists of our time, presents a mathematical model of evolution based upon existing mathematical theory and upon newer mathematical game theories. The "Hawk-Dove Game" illustrates predator-prey interactions in evolution. Chapter 3, "The War of Attrition," and chapter 8, "I. Asymmetric Games: Ownership," describe various species interactions such as territory, sex selection, and size selection. These

models are based upon the early work of physicists John von Neumann and O. Morgenstern.

___________. *Evolutionary Genetics*. New York: Oxford University Press, 1989.
This book, which is based upon the eminent author's lectures at the University of Sussex, is a summary of the genetics and molecular evolution of life. Chapter 3, "Evolution in Diploid Populations," and chapter 6, "Quantitative Genetics," present mathematical models of natural selection acting upon genes in organismal populations. These models include the important Castle-Hardy-Weinberg equations. Other major subject areas in this comprehensive book include the evolution of prokaryotes and eukaryotes, mitochondrial DNA, inbreeding, genetic drift, the evolution of sex, macroevolution, and the molecular evolution of hemoglobin in vertebrate blood. Numerous charts, reference lists, and diagrams support the text.

Spencer, Kevin C. *Chemical Mediation of Coevolution*. San Diego: Academic Press, 1988.
Spencer's book is an analysis of coevolutionary patterns between herbivorous insect species and plant species. These two groups have evolved together based upon their connectedness in the environmental food chain. Plants produce toxic insecticides to fend off insect predators. Some insect species, however, have adapted to these toxins in amazing ways, sometimes to their own benefit. Furthermore, the insect species activity also can benefit the plants. Spencer cites a considerable quantity of evolutionary data to support his points.

Swanson, Carl P., Timothy Merz, and William J. Young. *Cytogenetics: The Chromosome in Division, Inheritance, and Evolution*. Englewood Cliffs, N.J.: Prentice-Hall, 1967, 1981.
This reference book is a very comprehensive, but understandable, study of the chromosome (DNA) from species to species. Changes in chromosome structure and the organization of genes located on chromosomes probably have been principal instigators of species evolution. This book illustrates evolutionary change on the chromosomal level using a tremendous amount of experimental data. Excellent photographs, data tables, sketches, and reference lists support the authors' arguments.

Takahata, Nayouki, and James F. Crow, eds. *Population Biology of Genes and Molecules*. Tokyo: Baifukan, 1990.
This volume is a collection of twenty-one papers by leading evolutionary geneticists in honor of Motoo Kimura, the developer of the neutral theory of molecular evolution, a controversial theory that maintains that many genes have neutral adaptability but may become favored by natural selection after millions of years. The collected papers discuss various aspects of the theory and how it has

influenced evolutionary theory. The reader should have a strong mathematical background.

Terzaghi, E. A., A. S. Wilkins, and D. Penny, eds. *Molecular Evolution: An Annotated Reader*. Boston: Jones & Bartlett, 1984.
This book is an outstanding collection of classic papers in molecular evolutionary research and theory. Reprints of the actual research journal articles and their appropriate citations are included. The numerous papers cover a wide range of issues, including the origin of life, the evolution of gene-coding sequences, the regulation of gene expression, and repeated sequences in DNA. The scientific authors include a who's who of evolutionary biology, including such names as Roy J. Britten, D. E. Kohne, Lynn Margulis, Francis Crick, Walter Gilbert, Eric Davidson, Carl Woese, and Leslie Orgel.

Wilson, Edward O., et al. *Life on Earth*. Stamford, Conn.: Sinauer Associates, 1973.
This enormous volume covers the science of biology strongly from an evolutionary stance. The origin of life and the sequential development of new species over the four-billion-year history of life on earth are described in detail with supporting experimental data. The mechanisms of evolution are studied along with the genetic properties of species populations.

Chapter 3
EVOLUTIONARY CONTROVERSIES

Lamarckism and Alternative Evolutionary Theories

Ardrey, Robert. *The Social Contract: A Personal Inquiry into the Evolutionary Sources of Order and Disorder*. New York: Atheneum, 1970.
In this book, Ardrey looks at evolutionary theory and its implications on humanity. He presents a historical survey of Darwin's work and links it with the molecular biological breakthroughs of the late twentieth century (such as Watson and Crick's discovery of the structure of DNA). He also addresses resistance to evolutionary theory on the part of creationists and discusses various debates among evolutionists concerning evolutionary theory. He looks at the social behavior of fish, reptiles, birds, and mammals and examines violence in humans.

Corsi, Pietro. *The Age of Lamarck: Evolutionary Theories in France, 1790-1830*. Translated by Jonathan Mandelbaum. Berkeley: University of California Press, 1988.
Corsi's work is a historical survey of the development of early evolutionary theory during the early 1800's. Jean Baptiste de Lamarck propounded the evolution of life by the acquisition of acquired characteristics. Corsi examines Lamarck's scientific researches and viewpoints in chapter 2, "Lamarck: The First Synthesis." Scientific and public opinions of Lamarck's theory are covered in subsequent chapters. Chapter 6, "Developments and Problems in Lamarckian Doctrine," describes the scientific evidence against the inheritance of acquired characteristics.

Csanyi, Vilmos. *Evolutionary Systems and Society: A General Theory of Life, Mind, and Culture*. Durham, N.C.: Duke University Press, 1989.
In this intriguing work, biologist Csanyi discusses an approach to evolution that mirrors sociobiological evolutionary and ecological thought. He considers various key steps in the evolution of life on earth, including origin, molecular evolution, punctuated evolution in which catastrophes open up niches for occupation by surviving species and their descendants, neural evolution in animals, and cultural evolution in higher animals (such as mammals). He even extends the process of evolution to the development and continuous changes in various human societies. The environment that we create is merely an extension of our genetic makeup.

Dupre, John, ed. *The Latest on the Best: Essays on Evolution and Optimality*. Cambridge, Mass.: MIT Press, 1987.
This collection of philosophical and theoretical essays on the mechanisms of evolution examines the theory of optimality in which species adapt to changing environments with maximum expression of their available resources. Concepts

discussed in these papers include simple models of complex phenomena, animal behavior, altruism, human evolution, and the evolution of mind and culture. Contributors to the volume include Elliott Sober, J. M. Smith, and Roger N. Shepard.

Eldredge, Niles. *Time Frames: The Rethinking of Darwinian Evolution and the Theory of Punctuated Equilibrium*. New York: Simon & Schuster, 1985.
Eldredge considers a raging debate among evolutionists: the problem of whether or not evolution proceeds smoothly and gradually with a steady rate of extinction and appearance of new species (gradualistic evolution) or whether evolution occurs in short, rapid bursts in which many species die off, followed by the adaptive radiation of capitalistic species (punctuated equilibrium). Eldredge examines the geological record of the past one billion years, the fossil record of extinctions, construction of phylogenetic trees relating organisms, and the evolution of certain good "marker" species (such as trilobites) to debate these two viewpoints. There is considerable support for both theories, although recent evidence may tilt the advantage toward punctuated equilibrium.

Farley, John. *The Spontaneous Generation Controversy from Descartes to Oparin*. Baltimore: The Johns Hopkins University Press, 1974, 1977.
In this historical work, Farley examines the shift in scientific thinking about life from a vitalistic viewpoint to a mechanistic viewpoint during the past three hundred years of Western civilization. Up until the mid-1800's, there was a raging debate between both camps of scientists concerning the supposed phenomenon of abiogenesis (the spontaneous generation of life from nonliving matter). Farley traces the roots of the controversy up to and including Louis Pasteur's famous swan-neck flask experiment, which refuted spontaneous generation. The molecular revolution of the twentieth century is described along with Alexander I. Oparin's primordial soup hypothesis, an idea that requires a kind of spontaneous generation.

Godfrey, Laurie Rohde, ed. *What Darwin Began: Modern Darwinian and Non-Darwinian Perspectives on Evolution*. Boston: Allyn & Bacon, 1985.
This collection of philosophical essays by leading evolutionary scientists looks at the pros and cons of various debates concerning evolutionary theory and aspects of the theory. Discussed concepts include origins of life, the element of chance in evolution, breaks and discrepancies in the fossil record, punctuated equilibrium versus gradualistic evolution, and the question of human origins. Contributing authors include Stephen Jay Gould, Motoo Kimura, David M. Raup, and Niles Eldredge.

Gould, Stephen Jay. *Ontogeny and Phylogeny*. Cambridge, Mass.: Harvard University Press, 1977.

In this lengthy, provocative book, Gould provides his usual unique perspective upon the mysteries of evolution. At the same time, he pokes fun at those who misinterpret the meaning of evolution. He begins the book with a brief history of evolutionary history and controversies from 1859 to 1900 and then focuses on abuses in the teaching/treatment of evolution during the first half of the twentieth century, including its association with Freudian psychoanalysis, the Scopes trial, and racism. Toward the end of the book, Gould addresses major new hypotheses concerning the mechanisms of evolution, including the Britten and Davidson model of gene regulation in higher organisms.

Hull, David L. *The Metaphysics of Evolution*. Albany: State University of New York Press, 1989.
This fascinating philosophical essay discusses the history of biological evolutionary theory with some input from other scientific disciplines (such as physics). He discusses major evolutionary ideas from Lamarck to Darwin to Max Planck. He looks at the nature of science, the mechanisms of natural selection, special topics in evolutionary theory such as altruism, and problems in the construction of phylogenetic trees (the relatedness of different species). Controversial evolutionary ideas such as Edward O. Wilson's sociobiology theory are presented and debated as well. Hull's text is supported by a good reference section.

Jacob, François. *The Possible and the Actual*. Seattle: University of Washington Press, 1982.
In this philosophical work, Nobel laureate Jacob, codiscoverer of the lactose operon in the bacterium *Escherichia coli*, considers the methods and mistakes of modern science and evolutionary theory. Chapter 1, "Myths and Science," addresses the self-induced ignorance of many people toward science, as is evidenced by the silliness of many science fiction books and movies. Chapter 2, "Evolutionary Tinkering," describes the processes of evolution, breakthroughs in DNA research, and the promises/hazards of biotechnology.

Jagersten, Gosta. *Evolution of the Metazoan Life Cycle: A Comprehensive Theory*. New York: Academic Press, 1972.
In this interesting approach to the evolution of animal species, Professor Jagersten analyzes the comparative embryology, the comparative anatomy and physiology, and the development/differentiation of the animal body to produce a bilaterogastraea theory of animal evolution. According to this controversial view of evolution based upon some solid facts of developmental biology, most of animal evolution is based upon events that occur very early in animal body development. An example of a critical stage in development is bilateral (mirror image halves) or radial (spiral) cleavage patterns of development during the thousand-cell gastrula stage that occurs only a few days after conception in most animal species.

Le Conte, Joseph. *Evolution: Its Nature, Its Evidences, and Its Relation to Religious Thought*. New York: D. Appleton, 1897.

This unusual book, written by a University of California geologist, provides extensive evidence for the occurrence of evolution. Le Conte presents extensive fossil record data, comparative anatomical analyses between different species, comparative embryology, and biogeography to support the Darwinian-Wallace Theory of Evolution. He then shifts to metaphysics, where he considers "The Relation of Evolution to the Idea of the Christ" (chapter 8) and "The Relation of Evolution to the Problem of Evil" (chapter 9).

Lima-de-Faria, A. *Evolution Without Selection: Form and Function by Autoevolution*. New York: Elsevier, 1988.

This brilliant work by a Swedish cytogeneticist takes a radically new stance upon the mechanisms of evolution. Lima-de-Faria emphasizes the symmetry and self-organizing principles of living organisms and maintains that the evolution of organisms primarily is influenced by environmentally imposed changes upon organisms. These environmental changes can include elementary particles, light, and electric and gravitational fields. These environmental changes affect the chemistry of the cell to respond via gene regulation. Given the genetic variability of the organismal population, the species may or may not adjust to the changes.

Lovelock, James. *The Ages of Gaia: A Biography of Our Living Earth*. New York: W. W. Norton, 1988.

This excellent book represents an important challenge to modern scientific views of life on earth. Lovelock, a scientist and inventor who operates from his home, not the confines of academia, originally proposed with biochemist Lynn Margulis that all life on earth functions together as one superorganism, a single unit. This Gaia Hypothesis, named after a mythical goddess of the earth, has received much criticism. Lovelock, however, provides considerable evidence supporting the hypothesis. Furthermore, he challenges individuals to stand up against many outdated ideas of the current scientific establishment.

Lovtrup, Soren. *Darwinism: The Refutation of a Myth*. New York: Croom Helm, 1987.

In this book, Lovtrup, a University of Umea zoologist, performs a historical and philosophical analysis of Darwin, his theory, and currently accepted mechanisms of evolutionary change. He analyzes molecular, ecological, and paleontological evidence to propose a theory of macromutation that is very reminiscent of Motoo Kimura's neutral theory of evolution, where most genes are neither advantageous nor detrimental to the organism. He embraces Kimura's theory and even goes so far as to propound Lamarckian inheritance as deserving preeminence over Darwinian evolution.

Luria, Salvador E. *Life: The Unfinished Experiment*. New York: Charles Scribner's Sons, 1973.

This short, philosophical essay on the nature and meaning of life is written by an eminent evolutionary geneticist. He describes the mechanisms of evolution, the debates over various aspects of evolutionary change, the gene as the agent of evolutionary change, and the future evolution of humans. He places a special emphasis upon the central dogma of molecular biology: DNA (genes) encodes RNA, which encodes protein.

Margulis, Lynn. *Symbiosis in Cell Evolution*. San Francisco: W. H. Freeman, 1981.

In this important work, the distinguished University of Massachusetts at Amherst evolutionary biochemist describes the development of eukaryotic species (multicellular forms such as animals, plants, fungi, and protists) as being brought about by the symbiotic coexistence of single-celled prokaryotes similar to bacteria. Margulis proposed this controversial theory during the 1960's, a theory that received much opposition but has gained acceptance by the evolutionary science community. Margulis presents compelling evidence for her theory from studies of intracellular organelles, cellular biochemistry, DNA sequence data, and the fossil record. She cites an extensive amount of data in a very clear style to support her brilliant theory.

Plotkin, H. C., ed. *Learning, Development, and Culture: Essays in Evolutionary Epistemology*. New York: John Wiley & Sons, 1982.

This excellent collection of philosophical essays by leading evolutionary scientists and philosophers examines various points of contention between scientists over certain aspects of evolutionary theory and mechanisms. Items discussed include the definition and classification of species, behavioral evolution, cultural evolution, and teleology. Contributors include such eminent scientists as Ernst Mayr, Konrad Lorenz, Jean Piaget, Richard C. Lewontin, and Gilbert Gottlieb. The essays are well referenced for further research.

Reid, Robert G. B. *Evolutionary Theory: The Unfinished Synthesis*. Ithaca, N.Y.: Cornell University Press, 1985.

Biologist Reid looks at several key evolutionary debates from pre-Darwinian times to the present. He begins with a history of evolutionary theory and ideas, debates over vital forces within living organisms, Lamarckian inheritance of acquired characteristics, and the triumphant arguments for evolution by Sewall Wright, Thomas Henry Huxley, and others. He looks at major genetic breakthroughs of the twentieth century, including Thomas Hunt Morgan's demonstration that genes, the units of inheritance and evolution, reside on chromosomes. Chapter 15, "The Homeostasis Paradox," examines the recent proposal by dinosaur paleontologist Robert Bakker that the dinosaurs were probably warm-blooded, a concept that runs counter to modern reptile biology.

Ruse, Michael. *Taking Darwin Seriously: A Naturalistic Approach to Philosophy.* New York: Basil Blackwell, 1986.
This philosophical approach to evolution combines evolutionary theory with the social and cultural evolution of humans. Ruse looks at primate and human evolution from various perspectives. In terms of social bonding, there are many similarities. In language, chimpanzee verbal linguistic skills are virtually nonexistent despite several experimental attempts. The acquisition of American sign language by chimpanzees and gorillas has been very impressive. Ruse transfers human evolution from primate origins to the moral and ethical social evolution of modern human society.

Russell, Peter. *The Global Brain: Speculations on the Evolutionary Leap to Planetary Consciousness.* Los Angeles: Jeremy P. Tarcher, 1983.
Psychologist Russell looks at human evolution from primate origins to dominion of the planet, although he stresses the importance of our realizing that we are only part of a massive superorganism that occupies the entire planet. Every part of this superorganism must exist in harmony with all other parts. Chapters 3, "Hidden Orders in Evolution," 6, "The Emergence of a Social Superorganism," and 12, "Toward a High Synergy Society," all emphasize the human role within nature.

Schmalhausen, I. I. *Factors of Evolution: The Theory of Stabilizing Selection.* Chicago: University of Chicago Press, 1949, 1986.
This important book is written by the late Russian evolutionary biologist who propounded the New Evolutionary Synthesis of Dobzhansky at a time when Soviet biology was being ravaged by the Stalinist suppression of evolution. Schmalhausen describes the elements of evolutionary theory with great clarity. He includes discussions of mutation, natural selection mechanisms, hereditary variability within natural populations, panmixia, and adaptive morphological reactions to environmental stresses.

Sober, Elliott, ed. *Conceptual Issues in Evolutionary Biology: An Anthology.* Cambridge, Mass.: MIT Press, 1984.
Essays and evolutionary debates are expounded in this volume by leading scientists such as Richard C. Lewontin, Edward O. Wilson, John Maynard Smith, Ernst Mayr, Richard Levins, Steven Jay Gould, and Richard Dawkins. They debate various issues such as optimization of species adaptability in evolutionary change, other theories of adaptation, the definition of species, methods of classifying species, and artifacts of the evolutionary process. The essays employ scientific data from different scientific disciplines ranging from molecular genetics to paleontology.

Sober, Elliott. *Reconstructing the Past: Parsimony, Evolution, and Inference.* Cambridge, Mass.: MIT Press, 1988.
In this informative book, evolutionist Sober presents some problems in the interpretation of evolutionary evidence. He discusses primarily arguments over the construction of phylogenies, or evolutionary trees, that attempt to show the relative relatedness between different species. Such phylogenies are constructed based upon comparative anatomic and DNA/protein sequencing data, information that often is incomplete or subject to diverse interpretations. Using some mathematical descriptions, Sober describes the science of evolutionary cladistics. He provides many illustrations and cites important references.

Steele, E. J. *Somatic Selection and Adaptive Evolution: On the Inheritance of Acquired Characters.* Chicago: University of Chicago Press, 1979, 1981.
In this gem of a book, immunologist Steele reconsiders the supposedly refuted ideas of evolution proposed by Jean Baptiste de Lamarck concerning the inheritance of acquired characteristics by organisms. Steele presents this viewpoint from immunology, an intricately complex area of organismal physiology that deals with the organism's ability to fight infection. Antibody-producing lymphocyte cells of the animal immune system can produce antibodies against virtually any foreign object, a phenomenon that seemingly would be impossible given a limited quantity of DNA per lymphocyte even with gene rearrangements. Steele describes the Somatic Selective Hypothesis of immune cells and relates the hypothesis to infections by RNA tumor viruses.

Waddington, C. H. *The Evolution of an Evolutionist.* Ithaca, N.Y.: Cornell University Press, 1975.
In this book, the eminent evolutionary scientist discusses the various concepts and controversies within evolutionary theory. He intermixes a unique blend of objective scientific data and metaphysical speculation into some of his arguments. He describes and debates developmental evolution of higher organisms, the Lamarckian inheritance of acquired characteristics, species colonization of new habitats, mimicry in the coevolution of species, altruistic animal behavior, the catastrophe theory, and the evolution of humans and language.

Weber, Bruce H., David J. Depeur, and James D. Smith, eds. *Entropy, Information, and Evolution: New Perspectives on Physical and Biological Evolution.* Cambridge, Mass.: MIT Press, 1988.
In this exciting volume, leading evolutionists describe the evolution of the universe and of life from a thermodynamic energy viewpoint. Concepts described include cosmological evolution from the big bang theory to planetary formation, aspects of thermodynamics such as energy conversion and entropy (disorder), and the origin of ecosystems and energy/food webs in living organismal communities. The average person will find the writing to be very clear.

Wicker, Jeffrey S. *Evolution, Thermodynamics, and Information: Extending the Darwinian Program.* New York: Oxford University Press, 1987.

In this interesting, alternative approach to evolutionary theory, Wicker investigates the physical basis of life, namely energy. He presents living organisms as manifestations of energy, with the information energy content of the organism stored within the DNA molecule, and energy interconversions occurring via a host of enzyme-mediated chemical reactions. The order, complexity, and eventual entropic decay of life is described from thermodynamic principles. Wicker even considers energy relationships as a contributing factor to the origin of sexual reproduction in animals and plants.

Wolsky, Maria de Issekutz, and Alexander Wolsky. *The Mechanism of Evolution: A New Look at Old Ideas.* New York: S. Karger, 1976.

In this survey work, the Wolskys study the history of evolutionary theory during the last two centuries, specifically focusing upon major debates and issues of contention. They examine genes and the occurrence of mutation, selection for or against mutated genes in populations, and mechanisms of speciation. They consider the modern debate of microevolution versus macroevolution. They also summarize the principal ideas concerning the development of extensive nervous systems in higher animals and the evolution of behavior that resulted from higher nervous system development.

Creationism

Aw, S. E. *Chemical Evolution: An Examination of Current Ideas.* San Diego: Master Books (Creation Life Publishers), 1982.

Aw's book is a detailed scientific but very opinionated attack upon evolution and biochemical evidence for the origin of life on earth. He prepares a very objective attack upon the primordial soup hypothesis of Oparin and Haldane and upon experiments that supposedly support that hypothesis. He begins these criticisms in chapter 2, "Making Life's Building Blocks in the Laboratory." In chapter 5, "The Evolution of Proteins," he disputes the construction of phylogenetic trees for relating species based upon molecular similarities between species.

Berra, Tim M. *Evolution and the Myth of Creationism: A Basic Guide to the Facts in the Evolution Debate.* Stanford, Calif.: Stanford University Press, 1990.

This thorough discussion of the evolution-versus-creation-science debate was precipitated by creationist attempts in the 1980's to suppress the teaching of evolution in American public schools. Berra presents the case for evolution, in the process refuting arguments championed by the creationists. He cites impressive data from DNA and protein sequencing experiments, the fossil record, and studies of hominid evolution. Excellent visuals, data tables, and references.

Denton, Michael. *Evolution: A Theory in Crisis*. Bethesda, Md.: Adler & Adler, 1985.

Denton, an Australian molecular biologist, argues against Darwinian evolution using evidences from molecular biology, the one area of science that so far has yielded the most compelling proof for the occurrence of evolution. Chapter 12, "A Biochemical Echo of Typology," looks at DNA nucleotide and amino acid sequence data to argue that every species of a class has a number of differences approximately equal to each and every other species, thereby making evolutionary classification impossible. Denton also looks at problems in other scientific fields, including paleontology.

Desmond, Adrian J. *The Politics of Evolution: Morphology, Medicine, and Reform in Radical London*. Chicago: University of Chicago Press, 1990.

In this interesting and unusual history, Desmond describes the conflict between the elitist Royal College of Physicians/Royal College of Surgeons and dissenting "radical" physicians and scientists who established their own museums and schools. Emerging from this political conflict is evolutionary biology, which was used as a tool in the debates between the opposing groups. The dissenters embraced spontaneous generation of life and the inheritance of acquired characteristics within living organisms. The Royal Colleges, however, led by the crossover moderate scientist Richard Owen, prevailed with some concessions while supporting Darwinian evolution. Desmond's book clarifies some misinterpretations regarding certain scientists during this period.

Eldredge, Niles. *The Monkey Business: A Scientist Looks at Creationism*. New York: Simon & Schuster, 1982.

In this attack on creation science, biologist Eldredge emphasizes the senselessness and wastefulness of the evolution-versus-creationism debate, where American children and science in general suffer at the expense of a war of words. Chapters 2, "America and Science: Hens and Eggs or Cats and Dogs," and 7, "Creationism, Religion, and Politics," examine the public's perception of science and the true roots of the controversy.

Futuyma, Douglas J. *Science on Trial: The Case for Evolution*. New York: Pantheon Books, 1983.

Evolutionary scientist Futuyma attacks the scientific creationists in response to their early 1980's efforts to enact legislation for balanced coverage of creationism and evolution in American public schools. Futuyma refutes the arguments of creation science proponents by presenting considerable evidence supporting evolution, including fossil evidence, molecular biology, mutation studies, and human evolutionary studies. He addresses the impact of evolutionary theory upon society and the sociocultural role of evolution for humans.

Gallant, Roy A. *How Life Began: Creation Versus Evolution.* New York: Four Winds Press, 1975.

Gallant supports evolutionary theory in this consideration of both sides of the creation/evolution controversy. He begins his book with a psychological and anthropological study of the human mind in chapter 1, "On the Fringe of Intelligence." In chapter 6, he provides "The Evidence for Evolution." He examines arguments from both sides of the dispute in chapters 7, "The Creationist's Stand," and 8, "The Evolutionist's Stand."

Gish, Duane T. *Evolution? The Fossils Say No!* San Diego: Creation Life Publishers, 1979.

In this short paperback, which is marketed for distribution to public school students, biophysicist Gish attacks the paleontological evidence supporting evolution. His criticism centers around anomalies in the fossil record and relies upon philosophical arguments to question an evolutionary interpretation of fossils. He also relies upon biblical and philosophical quotations but does not present any objective data supporting a creationist model for life on earth.

____________. *Speculations and Experiments Related to Theories on the Origin of Life: A Critique.* San Diego: Institute for Creation Research, 1972.

In this attack on evolution and origin of life research, biochemist Gish makes a good objective attempt to dismiss various molecular evidences that support the occurrence of evolution. Gish examines amino acid polymerization, Alexander I. Oparin's coacervate theory for cell formation, Sidney Fox's "Proteinoid Microsphere" model for cell formation, and the Miller-Urey experiment, which apparently verifies the plausibility of Oparin's primordial soup hypothesis.

Gould, Stephen Jay. *Hen's Teeth and Horse's Toes: Further Reflections on Natural History.* New York: W. W. Norton, 1983.

In this book, the eminent paleontologist Gould attacks the Reverend Jerry Falwell's Moral Majority and scientific creationists for their efforts at restricting the teaching of evolution in the public schools. He describes various evolutionary concepts by using many clever and often humorous analogies. Chapter 2, "Nonmoral Nature," discusses the fallacies of "good" versus "evil" in biological organisms. Other chapters, including chapter 13, "What Happens to Bodies if Genes Act for Themselves?" talk about the very essence and mechanisms of life itself.

Kitcher, Philip. *Abusing Science: The Case Against Creationism.* Cambridge, Mass.: MIT Press, 1982.

In this strong attack on creation science, Kitcher assaults the pseudoarguments made by creationists to justify the Genesis account of creation as being scientific. He begins with a discussion of Darwinian evolutionary theory for the layperson.

He then picks apart creationist attacks on evolutionary theory, including criticisms of the fossil record and molecular biological data as well as criticisms of common philosophical arguments involving entropy (increasing disorder in the universe) and thermodynamics. He stresses that in no way does Darwinian evolution threaten religious liberty or the existence of God.

Larson, Edward J. *Trial and Error: The American Controversy over Creation and Evolution*. New York: Oxford University Press, 1989.
Larson, a historian and former counsel to several United States congressmen, describes the political basis of evolutionary controversies from the Scopes trial of 1925 to the present. He points out that much of the teaching in public schools is based upon public opinion. Furthermore, legislators and judges base many of their decisions upon public opinion, even in the face of scientific evidence. Evolution has not been accepted in American public schools until recently, and it still faces considerable opposition.

Lester, Lane P., and Raymond G. Bohlin, with a response by V. Elving Anderson. *The Natural Limits to Biological Change*. Grand Rapids, Mich.: Zondervan, 1984.
This attack upon evolution by three creation scientists is a fairly objective examination of molecular genetics and the fossil record. Lester, a professor at Liberty University in Lynchburg, Virginia, and Bohlin, who works with Probe Ministries International, discuss genetics, mitosis, the process of mutation, and neo-Darwinian natural selection. They attack evolutionary theory by debating the accuracy of the fossil record and interpretations of molecular DNA and protein sequencing experiments that attempt to compare distinct species.

McMullin, Ernan, ed. *Evolution and Creation*. Notre Dame, Ind.: University of Notre Dame Press, 1985.
This balanced work incorporates the views of scientists and philosophers from both sides of the creationism-evolution debate. Alternative viewpoints on creation science and evolution are debated by Francisco J. Ayala, William H. Austin, John Leslie, David Kelsey, and others. The included papers are objective and avoid the needless rhetoric of other works on the controversy. Evolution and creationism are considered from scientific and philosophical positions that tackle problems in paleontology, genetics, ecology, psychology, and sociology.

Montagu, Ashley, ed. *Science and Creationism*. New York: Oxford University Press, 1984.
This book is a compilation of essays presenting the scientific case for evolution as opposed to the major criticisms of evolution made by scientific creationists. The book begins with a treatment of the 1981 Arkansas law requiring the balanced coverage of both creationism and evolution in the public schools. It then moves to essays comparing and contrasting creationist arguments with scientific data.

Many prominent evolutionary scientists, including Stephen Jay Gould, Isaac Asimov, Sidney W. Fox, and Kenneth R. Miller, provide interesting and compelling essays.

Morris, Henry M. *The Troubled Waters of Evolution*. San Diego: Creation Life Publishers, 1975, 1982.
This work is a philosophical attack upon evolutionary theory by a creation scientist affiliated with the Institute for Creation Research. Morris concentrates upon a few problems with the available evolutionary evidence such as radioactive dating techniques and the age of the earth, gaps in the fossil record throughout geological time, and the classification of organismal species based upon comparative anatomy and physiology. He makes extensive arguments backed by the biblical model of creation.

Noble, Edmund. *Purposive Evolution: The Link Between Science and Religion*. New York: Henry Holt, 1926.
Noble's book is a philosophical unification of evolution and creation, a view that argues for the acceptance and interrelation of both ideas. He emphasizes the reality of heredity and the conceit of humanity in his writing. Chapter 5, "Is the Universe an Organism?" addresses consciousness for the entire universe and the inevitable evolution of life from the universe. Chapters 40 and 41, "Parallels Between Religion and Science," examine the interrelationship of the development of Christianity and the emergence of evolutionary thought. Other chapters deal with the ability of evolution to explain life and the nature of consciousness.

O'Toole, George Barry. *The Case Against Evolution*. New York: Macmillan, 1926.
In this book, the theologist, philosopher, and animal scientist O'Toole argues against the Darwinian theory of evolution by attacking various aspects of the existing evolutionary evidence. He attacks theories concerning the origin of life on earth, inaccuracies in the fossil record, interpretations of homologous anatomical structures (such as wings, legs, and fins) between various species, and the construction of phylogenies that attempt to relate various species. He also attacks the archaeological evidence for human origins and theories of abiogenesis (life from nonlife). He concludes his book with a study of the nature of the human body and soul.

Ruse, Michael, ed. *But Is It Science? The Philosophical Question in the Creation/Evolution Controversy*. Buffalo, N.Y.: Prometheus Books, 1988.
This four-hundred-page collection of essays pools the resources of numerous evolutionary scientists of both the nineteenth and the twentieth centuries to debate the merits of creationist attacks upon evolution. Argumentative papers by Thomas Henry Huxley, Francisco J. Ayala, Stephen Jay Gould, and Richard Dawkins focus upon the fallacies of the creationist arguments and misinterpretations concerning the meaning of evolution.

______________. *Darwinism Defended: A Guide to the Evolution Controversies.* Reading, Mass.: Addison-Wesley, 1982.
In this comprehensive summary of evolutionary controversies, Ruse examines the arguments presented by scientific creationists against evolution, sociobiology, and punctuated equilibrium versus gradualistic evolution. His account is both historical and scientific, covering evolutionary theory from Lamarck to the present. Ruse cites considerable evidence supporting evolution, including molecular, paleontological, and ecological studies.

Taylor, Gordon Rattray. *The Great Evolution Mystery.* New York: Harper & Row, 1983.
In his last book preceding his death, Taylor looks at some major problems facing modern synthetic evolutionary theory, including how new species originate, why evolution appears to proceed in spurts, and whether or not life has purpose. He concludes that scientific data supports evolution and refutes creationism. He maintains, however, that evolution may be only a small part of a grander scheme that permeates the entire universe and that so far has eluded science.

Velikovsky, Immanuel. *Earth in Upheaval.* Garden City, N.Y.: Doubleday, 1955.
The scientist and philosopher Velikovsky considers evolution with respect to the earth's geological record. He compares and contrasts gradualistic evolution as compared to catastrophic evolution. With respect to the latter, he examines earth geological processes such as volcanism, earthquakes, cometary impacts, continental drift, and the changing orbit of the earth around the sun over millions of years. He also considers catastrophes (such as the Deluge) recounted in religious writings, Darwinian evolution, and the process of mutation in living organisms.

Verbrugge, Magnus. *Alive: An Enquiry into the Origin and Meaning of Life.* Vallecito, Calif.: Ross House Books, 1984.
Verbrugge, a fellow of the Royal College of Surgeons of Canada, takes an objective, critical look at evolution, creationism, and the methods of science. In chapter 1, "A Matter of Life and Death," he studies the growth of scientific thought from the Renaissance to the present, a phenomenon unfortunately accompanied by an increasing meaninglessness of life. In chapter 4, "The Word of God and the Thoughts of Man," he examines the merits of creationism. He questions "How We Look at Things" in chapter 11. In chapter 14, he emphasizes that "Life Is Very Special." This book is an enjoyable experience.

Wilder-Smith, A. E. *The Natural Sciences Know Nothing of Evolution.* San Diego: Master Books (Creation Life Publishers), 1981.
In this book, a noted creationist debates evolutionary theory by attacking the major sources of evolutionary data: molecular genetics, the fossil record, radioactive dating of rocks to obtain the earth's estimated age, ecology, and

comparative anatomy. Chapter 1, "Evolution as a Scientific Pseudofact," and chapters 2 and 3, "Biogenesis by Chance (parts 1 and 2)," examine the presumed scientific evidence supporting evolution and the origin of life on earth. Wilder-Smith's book contains many semantic arguments with emphases upon biblical creation.

Wilson, David B., ed. *Did the Devil Make Darwin Do It? Modern Perspectives on the Creation-Evolution Controversy*. Ames: Iowa State University Press, 1983.
This collection of essays by leading philosophers and evolutionary biologists is a response to fundamentalist religious attacks upon the teaching of evolution in American public schools during the early 1980's. Included articles discuss the history of evolution-creationist controversies, evolutionary theory, cosmology, the origin of life, thermodynamic theory and entropy (a common creationist argument using the laws of physics), world religions, and judicial battles in Iowa and Arkansas. Contributors to this volume include John H. Wilson, Warren D. Dolphin, and Henry A. Campbell.

Wooldridge, Dean E. *The Machinery of Life*. New York: McGraw-Hill, 1966.
Wooldridge's book is a survey of the development and evolution of life on earth. Special emphasis is placed upon the physical and chemical factors that are necessary for life's origin, the Oparin-Haldane primordial soup hypothesis, and the formation of the first living molecules (DNA, protein, complex sugars, and so forth). The genetic code of deoxyribonucleic acid (DNA) and ribonucleic acid (RNA) is described in detail leading up to the formation of the first cells and the evolution of cells into higher multicellular plants and animals.

Wysong, R. L. *The Creation-Evolution Controversy: Implications, Methodology, and Survey of Evidence*. East Lansing, Mich.: Inquiry Press, 1976.
In this book, Wysong takes a very balanced study of the creationist and evolutionary views of life on earth. He discusses "The Controversy's Importance" in chapter 1. He further asks, "Can Laymen Question?" in chapter 2, a query that he answers in the affirmative. He discusses evolutionary views on the origins of DNA, protein, and other biological molecules. He considers the primordial environment of earth and experiments (such as Miller-Urey) aimed at understanding prelife conditions on earth. He more or less leaves the decision over this controversy to the reader.

Chapter 4
COSMOLOGY

The Big Bang Theory and Cosmic Evolution

Arp, Halton. *Quasars, Redshifts, and Controversies*. Berkeley, Calif.: Interstellar Media, 1987.

Arp, a prominent radio astronomer, presents his view that quasars are not the most distant objects in the universe, contrary to accepted cosmological theories. He presents a considerable amount of previously unpublished and published data to support his views. He proposes that quasars are ejecta from galaxies and that the Doppler cosmic redshift measurement scale is in error. He also discusses the sociology and politics of American science.

Asimov, Isaac. *Counting the Eons*. Garden City, N.Y.: Doubleday, 1983.

This book is a collection of Asimov's science fact articles and essays that previously appeared in *The Magazine of Fantasy and Science Fiction*. The short essays discuss a variety of topics such as stars, the universe, physical laws that control the universe, and unsolved problems in astronomy and astrophysics such as the asymmetry of matter and antimatter in the cosmos. Asimov's articles are both captivating and humorous.

___________. *The Universe: From Flat Earth to Quasar*. New York: Walker, 1966.

Asimov presents a simple, progressive discussion of the earth, planets, stars, galaxies, and other structures of the universe. He describes the life cycles of stars as well as the beginning and end of the universe. Within our solar system, he guides the reader through the various planets, their histories, and the probability of life occurring beyond earth. Good photographs and diagrams supplement the text.

Barnsley, Michael. *Fractals Everywhere*. Boston: Academic Press, 1988.

This introductory textbook to the fractal geometry of nature is derived from a college mathematics course taught by the author at the Georgia Institute of Technology. It is an outstanding, although very mathematical, analysis of the fractal design of linear and nonlinear events, structures, and processes throughout the universe. A fractal is a repeating pattern reiterated over and over again in many different processes and structures throughout the universe. Barnsley teaches the reader how to study such fractal structures and how to construct them on a computer using a variety of programs.

Bartusiak, Marcia. *Thursday's Universe*. New York: Times Books, 1986.
This popular work is a comprehensive survey of the great achievements of twentieth-century astronomy written by a physicist. The big bang theory, the evolution of stars and galaxies, the search for the unseen dark matter needed to close the universe, and other cosmological problems are described, along with the principal scientists behind these studies. Bartusiak makes complex topics very understandable for a general audience.

Bondi, Hermann. *The Universe at Large*. Garden City, N.Y.: Anchor Press, 1960.
Bondi's work is a short introduction to cosmology for the average person. He describes the evolution of stars and galaxies while interweaving an informative discussion of how the laws of physics affect the evolution of the universe. He also discusses astrophysical processes on earth, including the effect of the moon upon the earth's ocean tides, magnetic fields, and the Van Allen radiation belts surrounding the planet. Bondi, a leading astrophysicist, teaches with both clarity and authority.

Breuer, Reinhard. *The Anthropic Principle*. New York: Springer-Verlag, 1990.
Breuer, a German astrophysicist at the University of Tübingen, discusses the evolutionary history of the universe and the inevitable development of life in the universe in this exciting and informative account. He propounds the anthropic principle, an idea supported by many scientists and philosophers, which maintains that the structure and evolution of the universe requires the formation of life eventually. Consequently, the universe should be teeming with life. Breuer provides a clear presentation of basis physics and chemistry whose principles are responsible for the universe and for life.

Broms, Allan. *Our Emerging Universe*. Garden City, N.Y.: Doubleday, 1961.
Broms' survey of cosmology is a thorough, but simple, discussion of the universe's evolution. He proceeds step-by-step from the big bang theory of the beginning of the universe to the ignition of stars and galaxies to the formation/accretion of planetary systems around stars. With respect to our solar system, he considers each planet and its ability to support life. He discusses various aspects of the earth-moon system, including meteor and asteroid impacts as well as continental drift. The book finishes with a discussion of the universe's future. Excellent astrophotographs are provided.

Clarke, Arthur C. *Report on Planet Three and Other Speculations*. New York: Harper & Row, 1972.
In this interesting collection of essays, novelist and scientist Clarke discusses the principles of physics in layperson's terms and examines humanity's place within the universe. He discusses the methods of science, the dangers of pseudoscience, and the difficulties posed by our ever-accelerating technology, including our rapid

destruction of earth's natural resources. Later chapters explore the significance of the Apollo moon landings and the "myth of 2001," the latter referring to his famous novel and film describing humanity's first contact with extraterrestrials.

____________. *2001: A Space Odyssey*. New York: Signet, 1968.
Based upon the screenplay written by Clarke and director Stanley Kubrick for the MGM masterpiece, Clarke's novel is a panorama of the universe and humanity's place in it. He describes the evolution of humanity from apes, speculates upon the possibility of extraterrestrial influences in human evolution, and examines humanity's future if it chooses not to destroy itself. Clips from the movie, starring Keir Dullea and Gary Lockwood, are included in this best-selling work.

Cloud, Preston. *Cosmos, Earth, and Man: A Short History of the Universe*. New Haven, Conn.: Yale University Press, 1978.
This book is an outstanding four-part interplay of cosmic and biological evolution. Part 1, "Cosmos," deals with the evolution of matter, stars, planets, and galaxies. Part 2, "Earth," considers dramatic changes in our planet during its 4.6-billion-year history. Part 3, "Life," looks at the evolution of living organisms on earth during the past 3.8 billion years. Part 4, "Man," describes human evolution and out future as well as pollution, overpopulation, and our search for life on other worlds.

Corliss, William R., comp. *Mysterious Universe: A Handbook of Astronomical Anomalies*. Glen Arm, Md.: The Sourcebook Project, 1979.
This work is a compendium of collected research journal articles concerning astronomical phenomena. The covered subject areas include planets, moons, orbits, eclipses, meteors, galaxies, quasars, black holes, enigmatic objects, and cosmology. The continual evolution of the solar system is a major theme in these scientific articles.

Darling, David. *Deep Time*. New York: Delacorte Press, 1989.
This thought-provoking book is the story of a single proton's existence from the beginning of time (that is, the beginning of the universe) to the end of everything, from the theories of the big bang to the big crunch. In the process of telling this story, Darling very cleverly illustrates key points and concepts in physics and cosmology. He uses simple examples and analogies. The story includes the evolution of stars, galaxies, planets, and life, through all of which the proton exists at some time.

Eddington, Sir Arthur. *The Expanding Universe*. New York: Cambridge University Press, 1952.
This short survey book, written by one of the foremost astronomers of the

twentieth century, expounds the inflationary universe theory in which the universe is expanding with galaxies moving farther and farther away from each other. Based upon his public lectures, the book covers all topics with great clarity. Mathematical equations describing universal expansion and Doppler redshift spectra of galactic recessions (movement away from each other) are vividly explained.

Fabian, A. C., ed. *Origins*. New York: Cambridge University Press, 1989.
In this provocative work, astronomer Fabian assembles authoritative essays on all aspects of the evolution of the universe from such distinguished scientists and thinkers as John Maynard Smith and Martin Rees. Covered topics range from the big bang origin of the universe to the evolution of galaxies to the formation of the solar system to the evolution of life on earth to the development of intelligence in primates.

Feder, Jens. *Fractals*. New York: Plenum Press, 1988.
Feder, a Norwegian physicist, presents the concept of a fractal to the beginning student of physics. A fractal is a simple pattern repeated over and over again in many different structures and processes that appear to be unrelated. As such, fractals may help to explain much of the evolution of the universe as a whole. Feder introduces the reader to the mathematical modeling of fractal structures in such commonly encountered things as clouds and snowflakes. He presents numerous case studies of fractal research and lists several basic mathematical formulae upon which many fractal structures are based.

Ferris, Timothy. *Galaxies*. New York: Harrison House, 1987.
This very popular introductory astronomy book is a simple, yet excellent, introduction to the subject for the average person. It is clearly written and contains numerous beautiful astrophotographs. The oversize photographs of such celestial objects as nebulae, the Andromeda galaxy, and the Pleides create a sense of awe and wonder in the reader. Charts and maps of galactic distributions describe our current understanding of the evolution of the universe from the big bang theory to the present.

Feynman, Richard P. *QED: The Strange Theory of Light and Matter*. Princeton, N.J.: Princeton University Press, 1985.
In this exciting introduction to quantum physics for beginning students having some exposure to physics and chemistry, the late California Institute of Technology professor presents several lectures given to the general public. Feynman brilliantly describes the principal particles of matter and energy, including the massless photons of electromagnetic energy and the numerous types of quarks that compose protons, neutrons, and electrons in matter. He presents complex ideas with great clarity and links seemingly unrelated events together.

Gamow, George. *The Birth and Death of the Sun: Stellar Evolution and Subatomic Energy*. New York: New American Library, 1945.
Gamow's book is a simple discussion of stellar chemistry and cosmic evolution for the general public. He begins with a description of the atom, the transmutation (interconversion) of elements in stars, and the life cycles of stars such as our sun. He also discusses the formation of planets around stars, the evolution of life, and the future of the universe. Photographs and wonderful examples/illustrations are presented as well.

____________. *The Creation of the Universe*. New York: Viking, 1952, 1961.
Gamow, one of the preeminent theoretical physicists of the twentieth century, expounds his own big bang theory, which describes the origin of the universe as a fiery explosion arising out of virtually nothing. His writing style, as always, is aimed at a general audience. He describes the physical principles behind galactic formation and cosmic evolution. He uses interesting and humorous analogies to drive his points home. He provides excellent illustrations and data tables to support the text.

____________. *Mr. Tompkins in Wonderland, or Stories of c, G, and h*. New York: Cambridge University Press, 1964.
In this brilliantly unusual explanation of physics, cosmology, and the universe, Gamow describes basic physical properties of the cosmos through the misadventures of a bank clerk, Mr. Tompkins. Included among the topics are Newtonian gravity and Einsteinian relativity, where *c* represents the speed of light, *G* represents the gravitational constant of the universe, and *h* represents Planck's universal constant. The story ends with Mr. Tompkins getting his girl; they live happily ever after. An appendix to the book includes lectures by Gamow on cosmology and quantum space-time theories.

Gilmore, Gerald, Ivan R. King, and Pieter C. van der Kruit. *The Milky Way as a Galaxy*. Mill Valley, Calif.: University Science Books, 1990.
This informative reference work is a compilation of lectures by the authors on our galaxy, the Milky Way. From mythological histories to galactic structures, they provide the reader with a very comprehensive coverage of our spiral galactic home. Various types of stars, stellar dynamics, the missing mass problem of closed universe cosmologies, and experimental methodologies for studying the universe are included.

Glass, Leon, and Michael C. Mackey. *From Clocks to Chaos: The Rhythms of Life*. Princeton, N.J.: Princeton University Press, 1988.
In this fascinating, yet detailed, book, Glass and Mackey present the relatively new science of chaos, a branch of thermodynamic physics that attempts to explain

events in the universe as a constantly shifting series of events between order and disorder, between predictability and random chaos. They place special emphasis upon chaotic events within living organisms, including the evolution of populations, chaotic diseases, and chaotic states within living organisms such as cardiac arrhythmias in heart patients. The authors cite a large quantity of experimental evidence in this cutting-edge scientific treatise.

Gleick, James. *Chaos*. New York: Viking, 1987.
Gleick's best-selling general science work is a tremendous introduction to the science of chaos for the layperson. Chaos is a branch of thermodynamic physics and chemistry that pictures a universe where processes shift backwards and forwards between chaotic and orderly states at the slightest disturbance to the process or system. He discusses the recent history of the science and principal personages in its development, including meteorologist Edward Lorenz, fractal computer programmer Benoit Mandelbrot, and physicist Mitchell Feigenbaum. Chaotic systems, the butterfly effect, fractals, strange attractors, and self-similarity are among the many fascinating topics that Gleick addresses in this beautiful book, which can be read repeatedly with wonder.

Greenstein, George. *The Symbiotic Universe: Life and Mind in the Cosmos*. New York: William Morrow, 1988.
Greenstein's popular book for a general audience is a fascinating presentation of the universe, its evolution, and the physical principles by which the universe has operated for approximately fifteen billion years. He describes complicated concepts in physics such as quantum mechanics with considerable ease and great clarity. He argues that the principle of symbiosis, by which processes and entities share and benefit from each other, is essential for the universe to exist and that the universe could not exist without the evolution of intelligent life.

Hawking, Stephen W. *A Brief History of Time*. New York: Bantam Books, 1988.
This enormously popular best-seller, written by one of the greatest theoretical physicists of this century, is a straightforward discussion of universal evolution from the big bang inflationary beginning of the universe to several possible futures for the universe (such as open, closed, or "flat"). Hawking describes cosmology, black holes (his own specialty), and grand unified theories (GUTs), which are aimed at solidifying our understanding of how our universe operates.

Hawking, Stephen W., and Werner Israel, eds. *Three Hundred Years of Gravitation*. New York: Cambridge University Press, 1987.
This book is a collection of scientific essays commemorating the three hundredth anniversary of the publication of Sir Isaac Newton's laws of motion, the *Philosophiae Naturalis Principia Mathematica*. The essays cover our current knowledge

of the universe and cosmology. Special problems are discussed such as the unidentified dark matter in the universe. Contributors include many of the world's great physicists, including Hawking, Israel, Kip Thorne, Roger Penrose, Alexander Vilenkin, and Alexander Linde.

Hiley, B. J., and F. David Peat. *Quantum Implications: Essays in Honour of David Bohm*. New York: Routledge & Kegan Paul, 1987.
This book is a compendium of short scientific and philosophical essays that relate to the branch of physics called quantum mechanics and that acknowledge a major contributor to this field, Dr. David Bohm. Quantum mechanics is a fundamental physical theory that describes the structure of the universe and of living organisms. The essays are daring and seek to understand the very nature of existence. Contributors include such great thinkers as Bohm, Richard P. Feynman, and Maurice Wilkins.

Hoyle, Fred. *Astronomy*. Garden City, N.Y.: Doubleday, 1962.
This book is a simple introduction to astronomy written by one of the greatest astronomers of the twentieth century. It is a detailed history of the subject and the scientists who made major breakthroughs in our understanding of the evolution of the universe. Special focus is placed upon the important achievements of early twentieth-century astronomy, including the discoveries of galaxies, galactic distance measurements, and the expansion of the universe. The book contains outstanding photographs, illustrations, important astronomical data, and practical methods in astronomy.

Kippenhahn, Rudolf. *Light from the Depths of Time*. New York: Springer-Verlag, 1987.
This wonderful little book is a tremendous introduction to astronomy and cosmology for the general public. Kippenhahn, in a fashion very similar to that of George Gamow, simplifies many complex astronomical concepts in a very entertaining way. For example, he describes what the universe would look like to a two-dimensional being. Chapter 4, "Plumbing the Depths of the Milky Way," discusses our place within our fairly typical galaxy, one of several hundred billion galaxies within our universe.

____________. *One Hundred Billion Suns: The Birth, Life, and Death of the Stars*. New York: Basic Books, 1983.
In this beautifully illustrated, well-organized book, the astrophysicist Kippenhahn describes the origin and evolution of the universe. He describes how stars and galaxies are born and what happens to them over time spans of billions of years. Chapter 5, "The Life Story of the Sun," explains the history of our own star. Chapter 13, "Planets and Their Inhabitants," discusses Frank Drake's Project Ozma and the search for life elsewhere in the universe.

Kuhn, Thomas S. *The Copernican Revolution: Planetary Astronomy in the Development of Western Thought.* Cambridge, Mass.: Harvard University Press, 1966.
This historical account concerning the development of astronomy and cosmological theories during the past several hundred years concentrates primarily upon the works of Nikolaus Copernicus, Galileo Galilei, and Johannes Kepler. These individuals set the theoretical and experimental foundations for modern astronomy, astrophysics, and cosmology. Kuhn discusses the ancient "two-sphere" universe as compared to the "modern" universe of today, which was developed to a large extent by the works of these scientists.

Layzer, David. *Cosmogenesis: The Growth of Order in the Universe.* New York: Oxford University Press, 1989.
Harvard astrophysicist Layzer describes the emergence of order from chaos in the universe beginning with the big bang theory and then proceeding through cosmic inflation, the development of stars and galaxies, and culminating in biological evolution, the emergence of intelligence, and the creation of technology on earth. His book is philosophical in scope and aims to instruct the general reader in the ways of science, in the structure of the universe, and in how things work. He also proposes some important ideas, including a popular theory that the universe may have started with a very cool (such as absolute zero) slow expansion instead of the predominantly maintained primeval fireball, a new proposal that makes sense thermodynamically.

Leslie, John, ed. *Physical Cosmology and Philosophy.* New York: Macmillan, 1990.
This provocative volume is a collection of essays written by leading scientists and philosophers concerning the origin, nature, and future of the universe. Major topics include the existence of God and the anthropic principle, which maintains that the evolution of the universe inevitably leads to the development of life everywhere, and intelligent life at that. Contributors to this volume include Bondi, Stephen Jay Gould, George Gamow, and Richard Swinbourne.

Lightman, Alan. *Ancient Light: Our Changing View of the Universe.* Cambridge, Mass.: Harvard University Press, 1991.
In this enthralling book, Lightman, an MIT astrophysicist, takes the reader on a historical tour through the history of the universe, from the big bang origin of the universe to the evolution of elements, stars, galaxies, and life itself. He describes major events in the evolution of the universe. He also describes the history of the scientific development of our current understanding of the cosmos, beginning with the ancient Greeks and proceeding through to the grand unified theories (GUTs) of present-day astrophysics. He also speculates upon the anthropic principle, the idea that the universe is geared toward the inevitable development of life.

Lightman, Alan, and Roberta Brawer. *Origins: The Lives and Worlds of Modern Cosmologists*. Cambridge, Mass.: Harvard University Press, 1990.

This book is a unique look at cosmology and the evolution of everything through the eyes of the theoretical physicists and astronomers who paved the way for our improved understanding of the cosmos. Lightman and Brawer begin the book with a brief overview of cosmological principles and theories. Then they proceed to interviews with more than two dozen eminent cosmologists, including Arno Penzias, Robert Wilson, and Steven Weinberg. Major scientific breakthroughs, theories, and unresolved problems in astrophysics are discussed in detail.

Linde, Andrei. *Particle Physics and Inflationary Cosmology*. New York: Harwood, 1990.

This deep analysis of the physical subatomic principles behind the structural makeup and evolution of the universe is written by an internationally recognized Soviet astrophysicist. Linde describes the new inflationary cosmologies (such as universal expansion with some particles reaching supraluminal, faster-than-light velocities) within the context of particle physics (such as quarks and the string theory). He also describes space-time, quantum mechanics, and phase transitions within the primeval fireball in the early seconds of the big bang theory.

Merrill, Paul W. *Space Chemistry*. Ann Arbor: University of Michigan Press, 1963.

In this book, Merrill discusses the chemical makeup of planets, moons, stars, and other cosmic objects. The structure and properties of atoms, elements, and molecules are presented. Furthermore, a methodology known as spectroscopy is described; this important technique enables scientists to identify the chemical components of distant stars. He also considers other techniques such as instrument packages mounted aboard rockets for detecting the chemical components of stars and planets. Data tables and superb photographs support Merrill's clear writing style.

Meyers, Robert A., ed. *Encyclopedia of Astronomy and Astrophysics*. San Diego: Academic Press, 1989.

This comprehensive volume includes forty-one essays by leading astronomers and astrophysicists covering major topics and problems in astronomy and cosmology. Prime articles include "Cosmology," "Galactic Structure and Evolution," "Planetary Atmospheres," and several articles on planets, the sun and solar system, and stellar physics. Each article is supported by excellent photographs, data tables, and important reference lists.

Morrison, Philip, and Phylis Morrison. *Powers of Ten: About the Relative Size of Things in the Universe*. A *Scientific American* book. San Francisco: W. H. Freeman, 1982.

This interesting book, written by a prominent physicist/SETI researcher and his wife, is a proportional analysis of the sizes of objects in the universe based upon the metric, or logarithmic, scale of ten. From atoms to large molecules to viruses to bacteria to humans to continents to planets to stars to galaxies to the entire universe, the Morrisons plainly illustrate our place in the universe, however small or large it may appear from one's point of view. The book is packed with interesting photographs to convey size distributions between various objects.

Munitz, Milton K. *Theories of the Universe: From Babylonian Myth to Modern Science*. New York: Macmillan, 1957.
This lengthy work is a thorough discussion of the history of astronomical thought from ancient times to the present. The development of Copernican and Newtonian views of the universe are described in detail along with their further elaboration by the works of Kepler, Halley, and other scientists. The development of modern cosmology and the current view of the universe is described through the work of Edwin Hubble, George Gamow, and other great twentieth-century astronomers.

National Aeronautics and Space Administration. *Space Settlements: A Design Study*. Washington, D.C.: Author, 1977.
This attractive book is a discussion of spacecraft, machines, and living conditions needed for permanent human abodes in space. Numerous detailed designs of proposed space stations, space vehicles, and artificial environments are described both in the text and in beautiful artworks and diagrammatic plans. The book is a compelling argument for space exploration and the need for humans to plan the inevitable settlement of space.

Novikov, Igor. *Black Holes and the Universe*. New York: Cambridge University Press, 1990.
This short survey is an excellent presentation of big bang cosmology and black holes. Novikov is a leading Soviet theoretical astrophysicist and cosmologist whose expertise lies in black hole, or singularity, theory. Using clear and oftentimes funny examples, he describes the history of the universe from the big bang to the present. He also describes the predicted physical properties of black holes and their effects upon the possible futures of the evolution of the universe.

Ozima, Minoru. *Geohistory: Global Evolution of the Earth*. New York: Springer-Verlag, 1987.
In this short, attractive introduction to planetary geology for the beginner, Ozima describes the formation of our solar system from a solar nebula approximately five billion years ago, the accumulation of planetesimals to form the planets and their satellites, and the geological history of planet earth. Ozima describes the evolution of the earth's crust and mantle, including the activities of volcanism and plate tectonics. Ozima also addresses the thermodynamic processes of the earth's interior.

Reeves, Hubert. *Atoms of Silence: An Explanation of Cosmic Evolution*. Cambridge, Mass.: MIT Press, 1984.

Reeves's book is a beautiful exploration of the universe and humanity's place in it. Cosmic evolution is described along with the formation of stars, galaxies, and planets. He describes the origin, evolution, and future of life on earth. He also describes our future as a cosmic species that can astroengineer entire planets and solar systems. The element of chance and the physical evolution of the universe, however, eventually will outdo us.

Ringwood, A. E. *Origin of the Earth and Moon*. New York: Springer-Verlag, 1979.

In this exciting planetary geology book, Ringwood examines various theories describing the possible formation of the earth and moon. Ringwood concentrates upon the planetesimal bombardment during early solar system formation as the most probable explanation for the development of the planets and their satellites. *Apollo* lunar rock samples are compared with earth samples to elucidate their possible origins. The book is an excellent resource for earth and lunar data.

Rolfs, Claus E., and William S. Rodney. *Cauldrons in the Cosmos: Nuclear Astrophysics*. Chicago: University of Chicago Press, 1988.

This information-packed astronomy and astrophysics reference work is a detailed summary of current knowledge concerning stars and stellar processes. Chapter 1, "Astronomy—Observing the Universe," discusses quasars, various types of stars, galaxies, and cosmic evolution. Later chapters deal with the complex chemistry that occurs within stars, in particular the stepwise formation of heavier elements, the reason why the earth, other planets, and humans exist. The chemistry and mathematics are both detailed.

Ronan, Colin A. *Discovering the Universe: A History of Astronomy*. New York: Basic Books, 1971.

This simple introduction to astronomy summarizes major astrophysical principles and includes historical sketches of important astronomical breakthroughs. Chapter 5, "The Scale of Space," describes the organization of the universe. Chapter 14, "The Age and Origin of the Universe," and chapter 15, "Looking to the Future," describe the evolution of the universe.

Sagan, Carl, and Ann Druyan. *Comet*. New York: Random House, 1985.

In this book, the eminent astronomer Sagan and his wife describe the history, properties, and evolution of comets, objects that date to the beginning of our solar system and that occupy a significant proportion of the total mass of our planetary system. Chapter 12, "Mementos of Creation," describes the origin and evolution of comets. Chapter 8, "Poison Gas and Organic Matter," discusses the physical composition of comets. Some scientists believe that comets may be suitable for some forms of life.

Schatzman, E. L. *The Structure of the Universe*. Translated by Patrick Moore. New York: McGraw-Hill, 1968.

This compact reference book for the layperson is a well-organized description of the universe, its evolution, its structure, and its fate. The formation of galaxies and the types of galaxies are described. The text is supported by informative data tables, charts, illustrations, and beautiful astrophotographs. Schatzman discusses historical breakthroughs in astronomy and the laws of physics as they affect the universe, including matter and antimatter.

Seielstad, George A. *Cosmic Ecology: The View from the Outside In*. New York: Harcourt Brace Jovanovich, 1983.

Seielstad, assistant director of Green Bank, West Virginia, operations for the National Radio Astronomy Observatory (NRAO), describes the beauty and extent of the universe in this exciting book for the layperson. He leads the reader through the history of the universe and the major events that have occurred in its development. He speculates upon the evolution of life as a necessary part of the universe's evolution. He provides the reader with a clear description of humanity's place within the immense cosmos and stresses the importance for us to preserve the earth as humanity continues to explore space.

Silk, Joseph. *The Big Bang*. 1980. Rev. ed. San Francisco: W. H. Freeman, 1989.

This outstanding book is a comprehensive, up-to-date presentation of how we think the universe originated and evolved. The physical and chemical properties of the universe are described from microseconds after the big bang to the far distant cosmic future trillions of years from now. The properties and evolution of stars, quasars, galaxies, and black holes are described thoroughly but with great simplicity.

Singh, Jagjit. *Great Ideas and Theories of Modern Cosmology*. New York: Dover, 1961, 1970.

This enjoyable treatise on cosmic evolution sequentially describes such topics as the physics of the universe, the origin of the universe, the development of galaxies, and the life cycles of stars. The Doppler redshift spectra that illustrate cosmic inflation (expansion) are described along with influencing factors such as space, time, and gravity. The book ends with discussions of element formation within stars, planetary formation, and the development of life in the universe. Excellent illustrations and diagrams support the text.

Struve, Otto. *The Universe*. Cambridge, Mass.: MIT Press, 1962.

This book is another short synopsis of astronomy and cosmology by a leading astrophysicist. Struve explains galactic structure, evolution, and galaxy distributions in space with great clarity. He also describes stellar evolution and humanity's place in the universe (that is, space exploration). His book includes outstanding black-and-white astrophotographs, an adequate but not-too-difficult level of mathematics, and vivid illustrations. He also discusses such techniques as radio astronomy for viewing more remote portions of the universe.

Thiel, Rudolf. *And There Was Light: The Discovery of the Universe*. New York: Alfred A. Knopf, 1957.
Thiel's book is a lengthy historical survey of astronomical theories and thought from ancient times to the present. The evolution of our views on the evolution of the universe is described through the works of prominent astronomers such as Ptolemy, Copernicus, Galileo, Kepler, Halley, Hubble, and Eddington, among others. Book 5, "Rejuvenation," provides strong coverage of modern cosmological theories.

Toulmin, Stephen, and June Goodfield. *The Fabric of the Heavens: The Development of Astronomy and Dynamics*. New York: Harper & Row, 1961.
This work is another historical survey of major astronomers and their breakthroughs from the ancient Greeks and Egyptians to the present. Cosmology, the science of the universe and its origins, has changed over thousands of years due to the discoveries of many notable scientists, including Aristotle, Ptolemy, Nicolaus Copernicus, Galileo Galilei, and Johannes Kepler. The importance of Newtonian mechanics and laws of motion to our understanding of the universe is stressed throughout the book.

Verschuur, Gerrit L. *The Invisible Universe: The Story of Radio Astronomy*. New York: Springer-Verlag, 1974.
This interesting, short book is a history of radio astronomy, the extremely sensitive branch of astronomy that is enabling astronomers to see deeper into the universe's past based upon the radio signals naturally emitted by stars, galaxies, quasars, and many other cosmic objects. Chapter 3, "The Radio Sun and Planets," describes radio studies of our solar system. Chapter 7, "Interstellar Space: Is It Empty?" discusses some major problems in contemporary cosmology.

Zeilek, Michael, and Elske v. P. Smith. *Introductory Astronomy and Astrophysics*. 1973. Rev. ed. Philadelphia: Saunders College Publishing, 1987.
This information-packed book is an invaluable source of cosmic information for the serious amateur or professional astronomer. All aspects of astronomy and cosmology, ranging from stellar chemistry to the planets to galactic structure to the fate of the universe, are described in a very straightforward fashion. The mathematics is algebraic, but it can be followed with patience. The data tables, diagrams, illustrations, and photographs are superlative.

Preconditions for Life

Alexander, Jerome. *Life: Its Nature and Origin*. New York: Reinhold, 1948.
Despite its date of publication, Alexander's book is an good introduction to the origin of life question. He explores the molecular basis of life, including genetics, immunology, and the role of enzymes in many living organisms. Chapter 1, "How Did Life Originate?" includes a historical discussion of major scientists and their views on life's origins. In chapter 5, he asks, "What Are Living Units?" He concentrates upon the evolutionary change in genes and in the enzymes encoded by genes, upon which he elaborates in chapter 11, "Catalysis as the Efficient Cause of Evolution."

Angelo, Joseph A., Jr. *The Extraterrestrial Encyclopedia: Our Search for Life in Outer Space*. New York: Facts on File, 1985.
This reference work is an A-Z dictionary of terms, stars, spacecraft, and experiments related to the search for life off the earth. The descriptions of each item are very clear and thorough. An enormous amount of relevant scientific data is provided, mostly coming from astronomical sources, spacecraft design, lists of space probes launched to the moon and to other planets, and search efforts for detecting radio signals from extraterrestrial civilizations are included in this beautifully illustrated work.

Asimov, Isaac. *Environments out There*. New York: Abelard-Schuman, 1967.
Asimov's book, as is typical of his other writings, is a simple introduction to the solar system. He describes conditions on the Moon, Mars, Venus, Mercury, and the gas giants (Jupiter, Saturn, Uranus, and Neptune) based upon astronomical and space probe (such as the *Mariner* series) data. A table of important solar system data is supplied. Asimov also discusses search efforts for other planets beyond the orbit of Pluto.

____________. *The Wellsprings of Life*. New York: Abelard-Schuman, 1960.
As in all of his many works, the biochemist and writer Asimov provides an exciting and extremely informative introduction to the molecular basis of life on earth. He begins with "The Classification of Creatures" and "The Evolution of Evolution" in chapters 2 and 4, respectively. He then proceeds to the chemical basis of living cells, including "The Changing Chromosomes" (chapter 9) and "The Changing Gene" (chapter 10). In chapter 14, "The Living Molecule," he concentrates upon one of the central dogmas of molecular biology and evolution: DNA encodes RNA, which encodes protein.

Baugher, Joseph F. *The Space Age Solar System*. New York: John Wiley & Sons, 1988.
This lucid, up-to-date book can serve as an informative reference or leisure

reading for anyone interested in the solar system, its properties, and evolution. The sun, planets, moons of planets, asteroids, and comets are described in considerable detail. Excellent photographs, diagrams, and data tables are provided from the *Mariner*, *Apollo*, *Viking*, *Pioneer*, and *Voyager* space missions.

Bernal, J. D. *The Origin of Life*. New York: World Publishing, 1967.
In this detailed introduction to evolution and the origin of life, Bernal concentrates upon the primordial soup hypothesis of Alexander I. Oparin and J. B. S. Haldane. In chapter 4, he speculates upon "The Origin of Biological Membranes." Chapter 5, "Strange Turn of Biopoesis—The Origin of Biological Polymers," deals with a critical event in the formation of life on earth, the linking together of amino acids, simple sugars, and nucleotides to produce enormous, complicated molecules such as protein, carbohydrate, and nucleic acids, essential components for all life on earth. Appendices to the book include selections from both A. I. Oparin and J. B. S. Haldane on "The Origin of Life" and G. Mueller on "Carbonaceous Meteorites and the Origin of Life."

Beutner, R. *Life's Beginnings on the Earth*. London: Chapman & Hall, 1939.
Beutner's unique early look at the origins of life on earth may be dated, but it concentrates upon some basic physical mechanisms that are being espoused by leading evolutionary biochemists of today. In chapter 5 of part 1, he asks, "Are Crystals Produced by Vital Growth?" In part 2, he looks at the importance of polymerized compounds of carbon in all living organisms on earth and the possibilities that crystallization of such compounds from simpler molecules on the early earth may have contributed to the formation of life on earth. He also speculates on the possibility of life on other planets.

Buvet, R., and Cyril Ponnamperuma, eds. *Molecular Evolution I: Chemical Evolution and the Origin of Life*. New York: American Elsevier, 1971.
This book is a transcript of presentation papers made by leading evolutionary scientists at the second International Conference on the Origin of Life, held at Wakulla Springs, Florida, in 1970 under the auspices of NASA and the Florida State University Institute of Space Biosciences. Important scientific reports are made by Alexander I. Oparin, Fritz Lipmann, Lynn Margulis, B. M. Mednikov, M. Florkin, H. J. Morowitz, Cyril Ponnamperuma, A. A. Krasnovsky, M. S. Kritsky, N. V. Vasilyeva, R. W. Kaplan, and many others. Topics include early cells, polymerization of organic molecules, the distribution of organic substances in the universe, photochemistry, monomers, and theoretical problems in origin of life research.

Cairns-Smith, A. G. *Genetic Takeover and the Mineral Origin of Life*. New York: Cambridge University Press, 1982.
This outstanding book presents a dramatically alternate view to the mechanisms

of the origin of life on earth. Cairns-Smith maintains that the first living molecules, not cells, may have arisen from crystals of clay. He describes in detail evidence supporting the solid-state biochemistry of life and the properties of inorganic crystals. He outlines these views in chapter 5, "On the Nature of Primary Genetic Material." In chapter 8, he describes "The Entry of Carbon" and the subsequent evolution of DNA, RNA, proteins, lipid membranes, and cells. This book is an important contribution to origin of life research and is very clearly presented.

____________. *Seven Clues to the Origin of Life: A Scientific Detective Story*. New York: Cambridge University Press, 1985.
In this exciting, alternative view to the origin of life, Cairns-Smith maintains that life began as living molecules from inorganic crystals of clay. He makes distinctive arguments for this hypothesis beginning in chapter 3, "Build Your Own *E. coli*," where he discusses cellular components. In chapter 10, "Crystals," and chapter 11, "The Clay-Making Machine," he presents the evidence supporting his daring proposal. Chapter 13, "Evolving by Direct Action," describes how life may have developed from a crystalline matrix that preceded cells. Cairns-Smith makes simple, clear arguments.

Calvin, Melvin. *Chemical Evolution: Molecular Evolution Towards the Origin of Living Systems on the Earth and Elsewhere*. New York: Oxford University Press, 1969.
This speculative survey work on the origin of life is written by an authority in the field of biochemistry, Nobel laureate Calvin, discoverer of the photosynthetic light-independent reactions essential for the survival of all life on earth. He discusses the fossil record, hydrocarbon chemistry, and the relatively new science of molecular paleontology. He also thoroughly discusses prebiotic chemistry, the three-dimensional structures of molecules (such as DNA or proteins), the evolution of membranes and cells, and the search for extraterrestrial life. Calvin's work is loaded with scientific data from a variety of scientific disciplines such as astronomy, biology, chemistry, geology, and paleontology.

Carr, Michael H., R. Stephen Saunders, Robert G. Strom, and Don E. Wilhelms. *The Geology of the Terrestrial Planets*. Washington, D.C.: National Aeronautics and Space Administration, 1984.
This outstanding collection of scientific data from such NASA space missions as *Apollo*, *Viking*, and *Mariner* describes environmental conditions on Earth, the Moon, Mars, Venus, and Mercury. While concentrating upon the geological features of these planets, the described data provide the reader with strong information favoring or rejecting the possibility of life on these worlds. As such, it helps us to appreciate the conditions needed for life to occur on earth.

Crick, Francis. *Life Itself: Its Origin and Nature*. New York: Simon & Schuster, 1981.
In this stimulating book for the layperson, the eminent Nobel laureate and codiscoverer of the structure of DNA (with James Watson) presents an interesting discussion of life biochemistry, exobiology, and the methods of scientific inquiry. He describes the key molecular mechanisms of life in chapters 3, "The Uniformity of Biochemistry," and 5, "Nucleic Acids and Molecular Replication." He discusses prebiological conditions on "The Primitive Earth" in chapter 6 and examines "How Early Could Life Have Started?" in chapter 10. He considers the possibilities of life existing elsewhere in the universe in chapter 8, "Other Suitable Planets."

Day, William. *Genesis on Planet Earth: The Search for Life's Beginnings*. 2d ed. Foreword by Lynn Margulis. New Haven, Conn.: Yale University Press, 1984.
Day's book is a great scientific discussion of the origins of life on the early earth, although it can be detailed in certain parts. He begins with "Life Before the Precambrian" in chapter 3, followed by "The Age of Prokaryotes" in chapter 4. Chapters 8, "The Molecular Basis of Life," 16, "The Essentials of Life," and 22, "Particles of Life," speculate upon early events in the evolution of life and the nature of such life. He also examines the origins of cells, the prevalence of organic molecules in the universe, and the Gaia Hypothesis of James Lovelock and Lynn Margulis that maintains that all life on earth works together as a single unit.

Dole, Stephen H. *Habitable Planets for Man*. New York: Blaisdell, 1964.
Dole, a scientist with the RAND Corporation and a member of a U. S. Air Force team studying the possibility of humans living in outer space, presents the results of the study. Dole discusses scientific planetology, the study of the physical, chemical, and biological factors necessary for life to exist on other planets. A large quantity of still-relevant scientific data is provided, including the probabilities of certain nearby stars having inhabitable planets and the dynamics of stellar planetary system formation.

Dole, Stephen H., and Isaac Asimov. *Planets for Man*. New York: Random House, 1964.
This more general version of Dole's *Habitable Planets for Man* incorporates the expertise of biochemist Asimov. The book is a description of the conditions necessary for life to arise and exist in the universe and of the capacity for humans to live on other worlds. Based upon a RAND Corporation study, the book highlights planetary formation and the possibilities of life existing on other planets in our solar system. Also, several nearby stars such as Alpha Centauri A, Tau Ceti, 70 Ophiuchi A, and Delta Pavonis are considered for possible planetary systems.

Dyson, Freeman. *Origins of Life*. New York: Cambridge University Press, 1986.
This very brief work is, in the Dyson tradition, stimulation to the mind to consider all possibilities. The eminent physicist proposes that life may have originated first with cells (lipid membranes, nothing more) and then later with self-replicating molecules (DNA and RNA) that invaded these cells, thereby resulting in a symbiotic existence. This viewpoint virtually is the reverse of popularly held evolutionary theories. The book is based upon lectures given by Dyson at Trinity College, Cambridge, England.

Elsasser, Walter M. *The Physical Foundation of Biology: An Analytical Study*. New York: Pergamon Press, 1958.
This authoritative treatise on process and pattern within living organisms is both a physical and a philosophical description of life. The author, a theoretical physicist, considers several critical topics in biology, including homeostasis (the maintenance of constant conditions), feedback processes (such as hormones), information transfer (DNA to RNA to protein), and the storage of information within the mammalian brain (such as thinking, memory, intelligence). Elsasser also considers the living individual as an automaton, a programmed being, an idea first hypothesized by the American computer pioneer John von Neumann.

Folsome, Clair Edwin, introductions. *Life: Origin and Evolution, Readings from Scientific American*. San Francisco: W. H. Freeman, 1952-1979.
This collection of short summary articles by leading origin-of-life research scientists presents major experimental breakthroughs in such work for the layperson in the *Scientific American* tradition. Harold Urey describes "The Origin of the Earth" by accretion of smaller planetoids. George Wald discusses "The Origin of Life." Geoffrey Eglinton and Melvin Calvin provide gas chromatographic data from fossilized rock strata in "Chemical Fossils." Lynn Margulis presents her own discovery of "Symbiosis and Evolution." All of these *Scientific American* reprint articles contain excellent photographs, charts, and illustrative diagrams.

____________. *The Origin of Life: A Warm Little Pond*. San Francisco: W. H. Freeman, 1979.
In this book, Folsome, a University of Hawaii-Manoa professor, presents a solid introduction to the origins of the solar system and of life on earth in a short, concise format. He describes "The Formation of Planetary Systems" in chapter 3. He follows that with a discussion of the primordial earth in chapters 4, "Origin and Evolution of Terrestrial Atmospheres," and 5, "Chemical Evolution." He speculates upon early forms of life, both noncellular and cellular, in chapters 7, "Protobionts," 8, "Organic Automata," and 10, "Origin of the Genetic Protocell." He describes the famous Miller-Urey experiment and the results of Martian biochemical tests from the *Viking* missions.

Fox, Sidney W. *The Emergence of Life: Darwinian Evolution from the Inside*. New York: Basic Books, 1988.
This short, scientific, and beautiful book, written by one of the leaders in prelife biochemical research, is an exciting analysis of the beginning of cellular life on earth approximately 3.8 billion years ago. Chapters 4 and 5 describe "The Protocell: How It Looked" and "How the Protocell Behaved," respectively. In chapter 8, "The New Evolutionary Paradigm: Molecular Selection, Natural Selection, and Determinism," Fox considers recent hypotheses that the first living molecule may have been RNA, not DNA, long before the first cells evolved.

___________, ed. *The Origins of Prebiological Systems and of Their Molecular Matrices*. New York: Academic Press, 1965.
Fox, a leading expert on self-organizing systems such as life, has assembled scientific papers covering prebiotic chemistry on the early earth. Contributors to this volume include several of the foremost authorities in the field of life origin research, including Alexander I. Oparin, J. B. S. Haldane, Theodosius Dobzhansky, N. W. Pirie, Fritz A. Lipmann, A. E. Mirsky, F. H. Quimby, J. D. Bernal, E. Bradley, and M. S. Blois. Topics include the origins of genetic systems in early organisms, the polymerization of prelife molecules, and experimental simulations of early life environments.

Fox, Sidney W., and Klause Dose. *Molecular Evolution and the Origin of Life*. Foreword by Alexander I. Oparin. San Francisco: W. H. Freeman, 1972.
Fox and Dose's work is an authoritative survey of research into the origins of life on the early earth of four billion years ago. They cite a wealth of scientific data from paleontology, geology, biochemistry, and molecular genetics to support a framework describing events that led to the formation of life. They describe prebiotic (prelife) environmental conditions on the early earth. They also discuss the biochemistry of early life forms on the planet.

Greeley, Ronald. *Planetary Landscapes*. Boston: Allen & Unwin, 1985, 1987.
This information-packed book is a summary of all major solar system geological data, with the exception of the 1989 *Voyager* Neptune data. Written by a prominent NASA (National Aeronautics and Space Administration) and Arizona State University space geologist and based upon *Voyager* and *Viking* spacecraft data, the book provides the reader with a spectacular pictorial voyage through the planets of the solar system, from Mercury to Uranus. An outstanding text is intertwined with numerous breathtaking photographs. The most interesting data comes from the moons of Jupiter and Saturn, where several satellites have environmental conditions that potentially could support life.

Halvorson, Harlyn O., and K. E. Van Holde, eds. *The Origins of Life and Evolution*. New York: Alan R. Liss, 1980.

This tremendous work is a compilation of presentation papers made by leading origin-of-life researchers at a 1979 symposium held at the Marine Biological Laboratory in Woods Hole, Massachusetts. The outstanding topics include "Microspheres from the Swartkoppie Formation: A Review," by Paul K. Strother and Elso S. Barghoorn. K. E. Van Holde describes "The Origin of Life: A Thermodynamic Critique." Carl Woese presents "An Alternative to the Oparin View of the Primeval Sequence." Lynn Margulis examines the "Phyla for Bacteria." These articles represent short summaries of relevant data with good data tables, models, and references.

Hoyle, Fred, and Chandra Wickramasinghe. *Lifecloud: The Origin of Life in the Universe*. New York: Harper & Row, 1978.
In this controversial book, two prominent astronomers suggest that life may have originated in the universe first from an interstellar gas, the lifecloud. They base this hypothesis primarily upon the vast distribution of biochemical molecules in the interstellar gas or medium between the stars. They speculate upon the possibility of exotic life forms existing in interstellar molecular clouds. They begin these arguments in chapter 2, "Myths, Miracles, and the Origin of Life." They devote several chapters to the formation of stars and interstellar molecules.

Hutchison, Robert. *The Search for Our Beginning: An Enquiry, Based on Meteorite Research, into the Origin of Our Planet and of Life*. London: British Museum of Natural History/Oxford University Press, 1983.
In this short, clear text for the layperson, Hutchison describes the formation of the planet earth by accretion from particulate bombardment and the continual random collisions of solar system debris with the earth. Such collisions may have helped to direct the course of evolution on earth. Hutchison even goes so far as to speculate, as many scientists have done, that life may have been carried to earth via meteorites, since meteorites do contain organic prelife chemicals. In chapter 8, "Origin of the Earth and of Life—A Synthesis," he summarizes his viewpoints based upon considerable evidence.

Jackson, Francis, and Patrick Moore. *Life in the Universe*. New York: W. W. Norton, 1962.
This neat little book is a general synopsis of how life originated and developed on earth, plus how environmental conditions may favor or hinder the occurrence of life on other worlds. The authors describe the chemical basis of life on earth along with the molecular structures of key life compounds (such as DNA and amino acids). They describe the Oparin-Haldane primordial soup hypothesis and the possibility of life existing on other solar system planets and satellites. Diagrams and photographic plates illustrate life and its place in the universe.

Kenyon, Dean H., and Gary Steinman. *Biochemical Predestination*. New York: McGraw-Hill, 1969.
Kenyon and Steinman's authoritative work is a detailed chemistry of the origin of life. Their book begins with a history of the problem, the Oparin-Haldane primordial soup hypothesis, the Miller-Urey experiment, and other studies devoted to prelife chemistry. Considerable evidence supporting evolution is provided from several fields, including geology, chemistry, physics, astronomy, and biology. The authors provide an excellent explanation of the formation of such biopolymers as DNA, RNA, and protein. Extensive data tables and long reference lists make this work an invaluable origin-of-life reference.

Keosian, John. *The Origin of Life*. New York: Reinhold, 1964, 1968.
Keosian's very short book is a simple introduction to origin of life research. It begins with a discussion of evolution and controversies over the spontaneous generation of life, particularly with reference to Louis Pasteur, Frederick Engels, and Alexander I. Oparin. Chapter 4, "Abiotic Synthesis of Organic Compounds," discusses the primordial soup hypothesis of Oparin and Haldane as well as experiments supporting the hypothesis such as the famous Miller-Urey experiment that generated amino acids from an electrically activated gaseous mixture. Chapter 6 deals with "The Primitive Atmosphere and the Energy Factor" in the origin of life. Keosian also speculates upon the existence of extraterrestrial life.

Kimball, A. P., and J. Oro, eds. *Prebiotic and Biochemical Evolution*. New York: American Elsevier, 1971.
This book is a comprehensive assortment of presentation papers from two symposia on the origin of life: the 1968 University of Houston Symposium on Proteins and Nucleic Acids and the 1969 sixth meeting of the Federation of European Biochemical Societies in Madrid, Spain. The detailed papers cover topics such as the formation of coacervate drops, protocells, the primordial synthesis of pyrophosphates, the replication of early DNA sequences, and repetitive DNA in many organisms. Contributors include Alexander I. Oparin, Sidney W. Fox, Stanley L. Miller, Leslie E. Orgel, M. Parris, J. E. Sulston, Thomas H. Jukes, Roy J. Britten, and D. E. Kohne.

Kuppers, Bernd-Olaf. *Information and the Origin of Life*. Foreword by Carl Friedrich von Weizsacker. Cambridge, Mass.: MIT Press, 1990.
In this interesting theoretical work, Kuppers speculates upon the source of life's origins and suggests possible models for early macromolecular development and evolution leading up to the first cells and beyond. In chapter 1, he summarizes "The Darwinian Concept of Evolution." In chapter 6, "The Chance Hypothesis," he debates the prevalence of millions of species with regard to DNA mutational events over time. In chapter 9, "Random Sequences," he discusses Turing machines, computers, and artificial intelligence. He concludes this unusual work with chapter 12, "The Protosemantics of Biological Information."

____________. *Molecular Theory of Evolution: Outline of a Physico-Chemical Theory of the Origin of Life*. Translated by Paul Woolley. New York: Springer-Verlag, 1983.

This very mathematical work looks at the possible origins of life on earth and poses interesting models for how life might first have arisen and subsequently developed. In particular, Kuppers incorporates the idea of the hypercycle, a cycling interflow of information between living systems. In chapter 2, he describes the "Structure and Function of Biological Macromolecules." In chapters 4 and 8, he discusses "A Model System for Molecular Self-Organization" and "Self-Organization in Macromolecular Networks," respectively. He also examines possible origins of the genetic code in DNA.

Margulis, Lynn. *Early Life*. Boston: Science Books International, 1982.

In this exciting, simple book by an innovative evolutionary scientist, the development and evolution of life from its origin to the present is described. Chapter 2, "Life Without Oxygen," describes the first living molecules, the first cells, photosynthesis, and the first sexual reproduction between bacteria (conjugation). Chapters 3, "Life with Oxygen," and 4, "A New Kind of Cell," usher in the age of eukaryotic cells approximately one billion years ago. Margulis concludes the book with a description of the elaborate developmental processes, tissue specializations, and differentiation of cells in higher eukaryotes.

Mason, Stephen F. *Chemical Evolution: Origins of the Elements, Molecules, and Living Systems*. New York: Oxford University Press, 1991.

Mason's work is an exciting study of the chemical evolution of the universe, beginning with the formation of the light elements hydrogen and helium shortly after the big bang occurrence. With the condensation of stars and thermonucleosynthesis, hydrogen and helium are fused to produce heavier elements and molecules. The release of these substances from supernovae scatters them throughout interstellar space to become the components of planets, comets, and living organisms. Mason's book is a beautiful exploration of the universe and its chemical makeup.

Miller, Stanley L., and Leslie E. Orgel. *The Origins of Life on the Earth*. Englewood Cliffs, N.J.: Prentice-Hall, 1974.

This authoritative work on life's origins, written by two pioneer scientists in the field, provides a wealth of scientific information in a very clear, concise format. Miller and Orgel cite data from astronomical, geological, paleontological, chemical, and biological sources to outline a scenario for the formation and evolution of life on earth. They discuss the chemical conditions of the early primordial earth, Alexander I. Oparin's primordial soup, the Miller-Urey experiment, and the formation/evolution of cells. They include scientific data from their own and other scientists' breakthrough experiments. They also speculate upon the existence of life elsewhere in the universe.

Mueller, Richard. *Nemesis*. New York: Weidenfeld & Nicolson, 1988.
Mueller, a physicist, describes the asteroid impact theory of Walter and Louis Alvarez in this general, personal account of the two scientists and their controversial theory, which attempts to explain the extinction of the dinosaurs sixty-five million years ago. He speculates upon the possible existence of a companion star to the sun, a companion tentatively named "Nemesis," that periodically disturbs the Oort comet cloud and subsequently rains comets upon the inner solar system, including earth. Mueller emphasizes his own contributions to the theory in this entertaining work that illuminates the politics of science.

Oparin, Alexander I. *Genesis and the Evolutionary Development of Life*. Translated by Eleanor Maass. New York: Academic Press, 1968.
In this book, the eminent Russian biochemist reiterates his primordial soup hypothesis and presents major experiments to support the hypothesis. He discusses the chemistry of important life molecules such as ATP (adenosine triphosphate) and NAD (nicotinamide adenine dinucleotide), both of which are involved in energy metabolism, catalysis, and information storage in living cells. He elaborates upon prelife chemistry leading up to the first cells in chapters 3, "Formation of the 'Primordial Soup,'" 4, "Origin of Prebiological Systems," and 5, "Evolution of 'Protobionts' and the Origin of the First Organism."

___________. *Life: Its Nature, Origin, and Development*. Translated by Ann Synge. New York: Academic Press, 1961.
This short, concise work, written by the author of the primordial soup hypothesis for the origin of prelife chemicals, is a discussion of early life, the nature of life, and the evolution of life from single cells to complex eukaryotic cells. Chapter 1, "The Nature of Life," examines the physical properties and definition of living organisms and deals with such topics as cybernetics. Chapter 2, "The Origin of Life," looks at the formation of the solar system and the development of life on earth shortly after the earth's accretion. Chapter 3, "The Earliest Period of the Development of Life," examines the first cells, energy metabolism, and photosynthesis.

___________. *The Origin of Life*. Translated by Sergius Morgulis. New York: Dover, 1953.
This book is a cumulative summary of the Russian biochemist Alexander I. Oparin's ideas concerning the first living organisms on earth. Oparin describes the conditions of the early, primordial earth, the biochemistry of living cells, and the chemistry behind the formation of organic hydrocarbons essential for earth life. He proposes his famous primordial soup hypothesis, in which atmospheric

chemicals (such as hydrogen cyanide, carbon monoxide, ammonia, and nitrogen) react to produce organic molecules that rain down into the oceans, where they combine to produce the first cell. The book is authoritative, although it contains some errors due to the advance of scientific knowledge since the 1930's, the time of this book's original publication.

Oparin, Alexander I., and V. Fesenkov. *Life in the Universe.* New York: Twayne, 1961.
This speculative work is an early scientific analysis of the possibility of life arising elsewhere in the universe, particularly within our own solar system. Oparin and Fesenkov, the latter being the principal author, describe the structure of the universe, including stars, planets, galaxies, and quasars. They spend much of the book considering the possibility of the existence of life on Mars and/or Venus, questions that still to some degree remain unanswered.

Price, Charles C., ed. *Synthesis of Life.* Benchmark Paper in Organic Chemistry 1. Stroudsburg, Pa.: Dowden, Hutchinson & and Ross, 1974.
Price has assembled a stellar list of leading scientists and their already-published scientific papers describing the origins of life on earth. The reprints previously appeared in such scientific journals as *Nature*, *Science*, *Scientific American*, and *The Journal of Biological Chemistry*. Contributors to the volume include Stanley Miller, Cyril Ponnamperuma, Sidney W. Fox, Kaoru Harada, Fritz Lipmann, Sol Spiegelman, Arthur Kornberg, H. G. Khorana, Christian B. Anfinsen, George Wald, Masayasu Nomura, and Hans Kuhn. Topics include protein structure, DNA, optical activities of molecules, and ribosomes.

Raup, David M. *The Nemesis Affair: A Story of the Death of Dinosaurs and the Ways of Science.* New York: W. W. Norton, 1986.
Raup's book is an exciting and revealing look at the process of scientific discovery, in this case the Alvarez theory of dinosaur extinction by asteroid impact with the earth. Raup and Sepkoski followed Alvarez' theory with a computer analysis of extinctions over geological time, in the process discovering that there was an approximate twenty-six-million-year periodicity to massive extinctions of species. They speculated upon the possibility that the sun may have an unseen companion star, which they call "Nemesis." They discuss the controversy surrounding the theory, press coverage of the theory, and the politics of the American scientific establishment.

Rohlfing, Duane L., and Alexander I. Oparin, eds. *Molecular Evolution: Prebiological and Biological.* New York: Plenum Press, 1972.
To commemorate the birthday of origin of life scientist Sidney W. Fox, fifty-eight scientists contributed scientific and philosophical articles to this impressive volume. Oparin and Rohlfing offer the biography "Sidney W. Fox and Molecular

Evolution." John Keosian presents "The Origin of Life Problem—A Brief Critique." N. W. Pirie writes "On Recognizing Life." Alan W. Schwartz analyzes "The Sources of Phosphorus on the Primitive Earth—An Inquiry." J. C. Lacey, Jr., and D. W. Mullins, Jr., discuss "Proteins and Nucleic Acids in Prebiotic Evolution." Other contributors include J. Lawrence Fox, Thomas O. Fox, Ronald Forrest Fox, Linus Pauling, and Emile Zuckerkandl.

Rutten, M. G. *The Origin of Life by Natural Causes*. New York: American Elsevier, 1971.
This late professor of geology presents the geological case for the evolution of early life-forms nearly four billion years ago. In chapter 3, "Measuring Time in Geology," Rutten describes organic molecular evolution and radioactive dating methods for determining the age of fossils. Chapter 4, "The Biological Approach," looks at hostile conditions on the primordial earth, including the scarcity of oxygen in the atmosphere and the interconversion between inorganic and organic molecules. This attractive book contains much geological data, photographs, and maps.

Sagan, Carl. *The Cosmic Connection: An Extraterrestrial Perspective*. New York: Dell, 1973.
Sagan, in his inimitable style, presents a strong case for the exploration of space in this popular work. In so doing, he presents a brief history of the evolution of the universe, including the development of life and intelligence on an insignificant planet called earth. He considers the possibility of life existing elsewhere in the solar system, including Venus and Mars, and argues for attempts at radio communication with extraterrestrial civilizations. The book begins with an appropriate calling card from humanity that is mounted aboard the now-interstellar spacecraft *Pioneer 10*.

Scott, Andrew. *The Creation of Life: Past, Future, Alien*. New York: Basil Blackwell, 1986.
This simple, attractive book is an interesting historical and scientific study of the origin of life on earth. In chapter 4, "Stardust in a Pond," Scott describes the Miller-Urey experiment, which demonstrated that complex organic molecules could form from an energized gaseous mixture, thereby supporting the Oparin-Haldane primordial soup hypothesis. Chapter 8, "From Proteins to Paleobiologists," explores the possible origins of life and the evidence supporting various evolutionary models for the origin of life. Scott also contemplates extraterrestrial life.

Thaxton, Charles B., Walter L. Bradley, and Roger L. Olsen. *The Mystery of Life's Origin: Reassessing Current Theories*. New York: Philosophical Library, 1984.
In this critical examination of origin-of-life theory and research, the authors

challenge certain scientific stances concerning the mechanisms by which life may have first developed. They look at various modes of biochemical experimentation in chapter 3, "Simulation of Prebiotic Monomer Synthesis." In chapter 4, "The Myth of the Prebiotic Soup," the authors question whether life first arose in an ocean teeming with prelife chemicals or elsewhere. Chapter 10 examines "Protocells" and how the first cells may have formed.

Wolman, Yecheskel, ed. *Origin of Life*. Boston: D. Riedel, 1981.
This six-hundred-page volume contains detailed scientific papers presented by ninety-two scientists at the combined third meeting of the International Society for the Study of the Origin of Life and the sixth International Conference on the Origin of Life, held in Jerusalem in June, 1980. R. D. Brown discusses "Interstellar Molecules and the Origin of Life." J. Oro, C. Holzen, M. Rao, and T. G. Tornabene describe "Membrane Lipids and the Origin of Life." Other contributors include A. H. Delsemme, Michael D. Papagiannis, D. E. Ingmanson, Stanley Miller, M. J. Dowler, and J. E. van Trump.

Chapter 5
PRIMORDIAL LIFE

The First Cells

Alberts, Bruce, et al. *Molecular Biology of the Cell.* New York: Garland, 1983.
This work represents the combined efforts of six leaders in the field of molecular biology to summarize the major breakthroughs and trends in this explosive science. In chapter 1, they describe "The Evolution of the Cell," including speculations on events leading to the formation of the first cells nearly four billion years ago. Chapters 2, "Small Molecules, Energy, and Biosynthesis," and 6, "The Plasma Membrane," look at important aspects of cell function and structure. Other chapters are devoted to DNA, cell growth, cell division, and differentiation.

Bellairs, Ruth, Adam Curtis, and Graham Dunn, eds. *Cell Behavior: A Tribute to Michael Abercrombie.* New York: Cambridge University Press, 1982.
As a memorial to the late cell biologist Michael Abercrombie, this volume is a collection of papers dealing with the properties and functions of a variety of cells, both prokaryotic and eukaryotic. The contributed papers cover topics such as culturing cells for research, the extracellular matrix in between interacting cells, and mechanisms of motion of cells (such as traction, cilia, or flagella). Contributors include Albert K. Harris, J. M. Vasiliev, C. A. Middleton, Sally H. Zigmond, Peter B. Armstrong, Cheryll Tickle, James A. Weston, and J. G. Edwards.

Borisy, Gary D., Don W. Cleveland, and Douglas B. Murphy, eds. *Molecular Biology of the Cytoskeleton.* Cold Spring Harbor, N.Y.: Cold Spring Harbor Laboratory, 1984.
This five-hundred-page volume contains several dozen research papers presented at an April, 1984, symposium at the Cold Spring Harbor Biological Laboratory on Long Island, New York. The contributed papers discuss topics such as microtubules, microfilaments, cellular movement, and stability of the subcellular cytoskeleton. The approximately one hundred participants at the symposium include the editors, Susan Craig, Thomas E. Kreis, John V. Cox, Stephen R. Farmer, Henry F. Epstein, and Edward G. Fey.

Capaldi, Roderick A., ed. *Membrane Proteins and Their Interactions with Lipids.* New York: Marcel Dekker, 1977.
Ronald J. Baskin and Richard L. Jackson are among the many cellular biologists that have contributed to this important volume describing lipids, the organic molecules that double-up to form the lipid bilayers constituting the membranes forming all cells. The contributed articles deal with the lipid structures of cellular

membranes and of membranes within cells. Special interactions of enzymes with lipid membranes are also discussed, including ATPases and glycoproteins.

Cummings, Donald J., et al., eds. *Extrachromosomal DNA*. New York: Academic Press, 1979.
This volume contains presentation scientific papers from the fifteenth ICN-UCLA Symposium on Molecular Biology. Ruth Sager, Joseph S. Pagaro, Wesley M. Brown, Donald Cummings, Piet Borst, Igor Dawid, Sherman Weissman, and C. Fred Fox lead an impressive list of molecular biologists who contribute to this volume on extrachromosomal DNA, DNA molecules that are independent of nuclear DNA within living cells. Such DNAs include circular plasmids, transposable elements, and virus DNA. Major topics discussed in the papers include the evolution of mitochondrial DNA, chloroplast DNA, simian virus 40 (SV40), and eukaryotic extrachromosomal DNA molecules.

Davies, K. E., and A. P. Read. *Molecular Basis of Inherited Disease*. New York: Oxford University Press, 1989.
While this textbook is oriented toward the medical aspects of inherited genetic disorders, the authors clearly present in this informative guide the mechanisms by which mutations occur. Specifically, mutations occur by the alteration of DNA via chemical and/or radiation exposure. The damaged DNA encodes messenger RNA, which subsequently encodes a protein that may or may not function properly as a result of the mutation and corresponding amino acid changes.

Dawkins, Richard. *The Selfish Gene*. New York: Oxford University Press, 1989.
Dawkins' classic and controversial work correctly argues that evolution specifically acts at the level of the gene. The gene (DNA) encodes messenger RNA, which encodes protein, the visible, active traits of the organism. Dawkins goes to great lengths to argue that genes affect individual behaviors in the struggle for life. The ultimate goal is for the gene to replicate itself into the next generation via the organism in which it exists. Dawkins speculates on both selfish and altruistic genes in human and animal behavioral evolution.

Dixon, Bernard. *Magnificent Microbes*. New York: Atheneum, 1976.
In this simple introduction to microorganisms for a general audience, Dixon describes their prevalence and importance on earth, their origins and evolution, and their practical uses by humans. He emphasizes the role of microorganisms as scavengers within the recycling of energy throughout all living earth ecosystems. He examines their molecular constitution, including their genetics and the biochemistry of deoxyribonucleic acid (DNA). He addresses their roles in food production and in the causes of various diseases.

Drake, John W., and Robert E. Koch. *Mutagenesis*. Benchmark Paper in Genetics 4. Stroudsburg, Pa.: Dowden, Hutchinson & Ross, 1976.

These two University of Illinois, Champaign-Urbana professors have assembled a collection of famous research papers devoted to mutagenesis, the occurrence of mutations, and its effects upon DNA and cellular evolution. The reprinted papers are by such eminent geneticists as James Watson, Francis Crick, John W. Drake, George Streisinger, Sidney Brenner, and H. J. Mueller. The various types of mutations in DNA are described, including both frameshift and missense mutations.

Garrod, Archibald E. *Garrod's Inborn Errors in Disease*. Edited by Charles R. Scriver and Barton Childs. Foreword by Joshua Lederberg. New York: Oxford University Press, 1989.

Garrod was the first scientist to propose that certain human diseases are caused by mutation and, therefore, have a genetic basis. This book, a reprint of his 1931 book, is one of Garrod's last works. In it, he argues convincingly and correctly that different individuals have distinct genetic makeups and, therefore, have different susceptibilities to various diseases. These differences have arisen due to mutations in DNA over hundreds of millions of years of organismal evolution on earth.

Gennis, Robert B. *Biomembranes: Molecular Structure and Function*. New York: Springer-Verlag, 1989.

Gennis' impressive work is an in-depth analysis of lipids (fats) and how they combine to produce the cell membrane that essentially holds in the cell's contents, folds up to produce specialized subcellular structures, and houses critical catalytic enzymes for various cellular functions. Gennis describes the physical properties of lipids in relationship to a watery environment, the formation of lipid bilayers to produce membranes, and the dynamics of molecular transport across membranes. He provides excellent microphotographs, X-ray diffraction results, and other data to support the text.

Grobstein, Clifford. *A Double Image of the Double Helix: The Recombinant DNA Debate*. San Francisco: W. H. Freeman, 1979.

In this important book, Grobstein discusses the scientific importance and ethical problems involved in biotechnology, the branch of molecular biology and genetics that has helped scientists to understand the nature of life but that also has opened up new questions concerning human manipulation of life. He describes research into DNA, the information molecule for all living organisms, along with policies and regulations affecting cloning and other related genetic research. Grobstein's approach to the subject is very scholarly.

Guschlbauer, W. *Nucleic Acid Structure: An Introduction*. New York: Springer-Verlag, 1976.

In this detailed scientific presentation, Guschlbauer presents the biochemical properties of DNA, the information organic molecule in all living organisms. Guschlbauer describes the Watson-Crick "B" form of DNA, alternative forms of DNA, the genetic code, gene regulation mechanisms within the cell, mutation, and how DNA is used by the cell to encode enzymes and other proteins that essentially make the cell the living entity that it is. A wealth of experimental data is cited.

Hames, B. D., and D. M. Glover. *Gene Rearrangement*. New York: Oxford University Press, 1990.

Douglas E. Berg, Lexi J. T. van der Ploeg, George R. Stark, and colleagues describe the mechanisms by which segments of genetic information (DNA) are exchanged between different chromosomes and between different cells in both prokaryotic and eukaryotic species. Gene rearrangement has played a critical role in the evolution of life on earth. The contributing authors examine the causes and effects of such gene rearrangements. They also look at the medical aspects of such rearrangements in certain pathogenic species such as the trypanosome, the cause of sleeping sickness.

Inouye, Masayori, and Bernard S. Dudock, eds. *Molecular Biology of RNA: New Perspectives*. San Diego: Academic Press, 1987.

Inouye and Dudock have assembled an impressive collection of research and survey papers from a stellar list of expert scientists who work with ribonucleic acid (RNA), the other nucleic acid that usually receives second billing to DNA. Several evolutionary scientists believe that RNA may have been the first living form on earth. Contributors to this volume include Nobel laureates Thomas Cech and Sidney Altmann, who discuss their discoveries of autocatalytic RNA in *Tetrahymena* and *Escherichia coli*, respectively. Other contributors include Joan Steitz, Roy Parker, Bruce Patterson, Tom Maniatis, and Jun-Ichi Tomizawa. Gene regulation, RNA viruses, introns, exons, and antisense RNA are among the many other important topics discussed in this exciting book.

Jukes, Thomas H. *Molecules and Evolution*. New York: Columbia University Press, 1966.

Jukes's short, concise book describes the subcellular roles of various important molecules and the evolution of cells via evolutionary changes within these molecules. Jukes discusses DNA and the genetic code, the transmission of genetic information from DNA to RNA to protein, the effects of mutations upon the cell, and the microbiology of various prokaryotic cells. Excellent illustrations and photographs support his writing.

Karp, Gerald. *Cell Biology*. New York: McGraw-Hill. 1979.
Karp's work is an excellent survey of the cell for the beginning biology student. Clear writing, descriptive summaries, microphotographs, and illustrations help the reader to understand the intricate complexity of cellular biochemistry and structure. Karp sequentially describes the protein enzymes that control many cellular events, the lipid membranes that form the cellular periphery, cellular metabolism, gene arrangement and control, and cell division. His book includes special chapters describing the immune system, cancer, and aging in eukaryotic organisms.

Kornberg, Arthur, and Tania A. Baker. *DNA Replication*. San Francisco: W. H. Freeman, 1974, 1980, 1992.
Nobel laureate molecular biochemist Kornberg and associate Baker have in this mammoth volume compiled the most thorough summary of DNA, its structure, its copying mechanism within reproducing cells, and its gene regulation within all life on earth. This detailed reference is required reading for the serious geneticist and molecular biologist. Major breakthroughs in DNA biochemistry, cloning, and gene regulation are summarized in this tremendous work.

Lewin, Benjamin. *Genes*. New York: Oxford University Press/*Cell* Press, 1983, 1985, 1987, 1990.
Lewin, a prominent molecular biologist and editor of the prestigious research journal *Cell*, discusses up-to-date knowledge concerning the nature of genetic information (DNA) and the specific gene control mechanisms that are the focus of intensive research in both prokaryotic and eukaryotic cells. Lewin addresses the basic structure and functioning of the gene in the first several chapters. He also tackles the processes of mutation within genes in subsequent chapters. His book is a definitive survey of the field and is an invaluable reference source.

Li, Wen-Hsiung and Dan Graur. *Fundamentals of Molecular Evolution*. Sunderland, Mass.: Sinauer Associates, 1991.
In this fascinating book of molecular evolutionary biology, Li and Graur, two leading experts on molecular evolutionary theory, describe the methods of examining molecules from different species to determine the evolutionary relationship between these species. The authors describe comparative DNA and protein sequencing between different species of animals, plants, fungi, protists, and bacteria. They also discuss problems in interpreting data and in constructing possible phylogenetic trees relating different species.

Locke, David M. *Enzymes: The Agents of Life*. New York: Crown, 1989.
In this general introduction to basic biochemistry, Locke describes enzymes, the protein molecules within cells that catalyze (speed up) essential chemical reactions for life. Enzymes are encoded by specific genes (DNA) within cells; conse-

quently, Locke addresses the effects of mutations upon cellular genes via their effects upon enzymes. Mutant enzymes that function improperly may threaten the cell's survival or may contribute to the cell's further evolution. Locke also addresses the uses of enzymes in industry, nutrition, and medicine.

Losick, Richard, and M. Chamberlin, eds. *RNA Polymerase*. Cold Spring Harbor, N.Y.: Cold Spring Harbor Laboratory, 1976.
Walter Gilbert, Arthur Kornberg, D. M. Rothstein, and Samuel B. Weiss highlight an impressive list of scientists who attended a 1975 Cold Spring Harbor meeting on RNA and who contributed papers to this nine-hundred-page summary of ribonucleic acid and RNA polymerase, the protein enzyme that helps to synthesize RNA. The detailed scientific papers discuss the genetic control of DNA, the reading of the genetic code in DNA by RNA polymerase, and the manufacture of RNA from cellular components, which eventually leads to the manufacture of protein by ribosomes. The intricate cellular biochemistry that is responsible for life itself is presented in this volume.

McPherson, M. J., ed. *Directed Mutagenesis: A Practical Approach*. New York: Oxford University Press, 1991.
McPherson has assembled several important papers by leading geneticists to describe how mutations occur in DNA (genes) and how scientists are using this knowledge to deliberately mutate genes in various organisms. Such deliberate mutations enable scientists to understand how such genes function and how the protein products of these genes affect the organism. Contributing authors include S. Inouye, P. G. Stockley, and J. R. Knowles. Current mutagenesis protocols are included in the various papers.

Nystrom, Richard A. *Membrane Physiology*. Englewood Cliffs, N.J.: Prentice-Hall, 1973.
Nystrom, a University of Delaware scientist, describes in detail the makeup and functioning of lipid membranes, molecules that structurally make the cell what it is. He uses chemical studies and electron microphotographs of cellular membranes to demonstrate their properties. He presents popular models of membrane functioning, including the Peter Mitchell chemiosmotic hypothesis. The energy molecule adenosine triphosphate (ATP), cation transport across membranes, active transport across membranes, and "electric" organs on membranes are considered.

Prescott, David M. *Cells: Principles of Molecular Structure and Function*. Boston: Jones & Bartlett, 1988.
Prescott, a biologist at the University of Colorado at Boulder, presents the general biology student with an introduction of cell biology and genetics. He presents the basic properties, physical structure, and functions of subcellular structures for both prokaryotic and eukaryotic cells. He also deals with genetics, inheritance,

gene regulation during development, evolution of cells, cell motility, and cellular reproduction.

Schulz, G. E., and R. H. Schirmer. *Principles of Protein Structure*. New York: Springer-Verlag, 1979.
This outstanding discussion of proteins, the organic molecules that catalyze (speed up) chemical reactions and provide structure to cells and to entire organisms, is written by two leading authorities on the subject. Schulz and Schirmer describe the twenty basic amino acids found within the proteins of all living organisms on earth, the dehydration biosynthetic linkage of these amino acids to form peptide bonds and corresponding proteins, the primary, secondary, and tertiary structures of proteins, and how proteins function within living cells.

Stern, Herbert, and David L. Nurney. *The Biology of Cells*. New York: John Wiley & Sons, 1965.
In this book, the basic properties and functions of cells are described in detail from genetics to cytoskeletal structure. Chapter 8, "The Acquisition of Oxidative Energy: Electron Transport," describes cellular respiration within both heterotrophic and autotrophic cells. Chapters 12, "Molecular Replication," and 13, "Molecular Transcription," describe the behavior and properties of cellular DNA in prokaryotic and eukaryotic cells. The biochemical principles of cellular processes are well represented.

Stryer, Lubert. *Biochemistry*. 3d ed. San Francisco: W. H. Freeman, 1988.
Stryer's discussion of biochemistry is a tremendously outstanding introduction to the subject for the layperson as well as an excellent reference source for the expert biologist. Stryer describes the chemical makeup of the cell and of viruses, in the process providing very clear descriptions accompanied by beautiful illustrations and microphotographs. The biochemistries of DNA, RNA, protein, lipids, sugars, cellular metabolism, and cellular division are discussed in detail.

Szekely, Maria. *From DNA to Protein: The Transfer of Genetic Information*. New York: John Wiley & Sons, 1980.
In this beautiful work, Szekely describes the fundamental biosynthetic pathway critical to all life on earth and that essentially defines life on earth, the pathway from DNA to RNA to protein. She describes the replication of DNA, the copying of chromosomes within cells prior to cellular reproduction. She describes the chemical basis of transcription, the process by which the genetic code of DNA is used to encode RNA, and the chemical basis of translation, the process by which this RNA is used to encode protein, the actual active molecule within cells.

Thomas, Lewis. *The Lives of a Cell: Notes of a Biology Watcher*. New York: Viking Press, 1974.

In this popular bestseller, Thomas, president of the Sloan-Kettering Memorial Cancer Center, describes for a general audience the inner workings of the basic unit of life: the cell. He describes the molecular composition and interconnections within the cell with great simplicity. He studies the role of certain microorganisms in disease processes, the roles of microorganisms in human technology and medicine, and the development of computers as a new step in the evolution of life on earth. He examines the workings of science and argues that basic research will yield promising technological breakthroughs in the coming decades.

Thompson, E. Brad, and John Papaconstantinow, eds. *DNA, Protein Interactions, and Gene Regulation*. Austin: University of Texas Press, 1987.
Jeremey M. Berg, Alexandra Glucksmann, and Leonard A. Zwelling are among the many contributors to this volume describing the molecular basis of gene regulation, the process by which genes (segments of DNA) are turned on or off, thereby affecting the functioning and development of the cell in which the change has occurred. Proteins and hormones are demonstrated to play important roles in allowing or suppressing the control regions on genes, in so doing affecting the ability of the enzyme RNA polymerase to read genes and to encode RNA leading to protein production.

Vol'kenshtein, Mikhail V. *Molecules of Life: An Introduction to Molecular Biology*. Translated by Serge N. Timasheff. New York: Plenum Press, 1970.
This detailed survey of molecular biology, cell biology, and genetics is very comprehensive in scope. Vol'kenshtein approaches cell structure and function from a physical standpoint, in so doing emphasizing the physics of cellular events. In chapter 1, he describes the relationship between "Physics and Biology." In chapter 2, he discusses the "Physics of Macromolecules." DNA, gene control, protein synthesis, and polymerization of macromolecules within the cell are described in detail.

Watson, James D., John Tooze, and David T. Kurtz. *Recombinant DNA: A Short Course*. A *Scientific American* book. New York: W. H. Freeman, 1983.
Nobel laureate Watson, director of the prestigious Cold Spring Harbor Laboratory, and his colleagues herein describe the principles of cloning and molecular biological research along with valuable techniques for conducting successful experiments. The book is clearly written and is accessible to the layperson. Many basic properties of cellular biochemistry are presented, particularly for prokaryotes and viruses, since prokaryotes and viruses are the primary tools for biotechnology research. Many aspects of molecular genetics research are described in detail.

Zubay, Geoffrey. *Biochemistry*. Reading, Mass.: Addison-Wesley, 1983.
Zubay's survey of cellular biochemistry is an extremely detailed, yet excellent, introduction to the subject for the serious student of biology. The biochemical

basis of life is described in intricate detail. Early chapters are devoted to the kinetics of enzyme catalysis and the biochemistry of cellular respiration within eukaryotic and prokaryotic organisms. Later chapters approach the complexities of gene regulation and hormones.

Mutation and Selection

Anisimov, Vladimir. *Carcinogenesis and Aging*. Vol. 1. Boca Raton, Fla.: CRC Press, 1987.
Anisimov's book is an outstanding study of the physical and chemical processes that may be responsible for aging and cancer. He treats the various theories of aging, all of which implicate mutational events in DNA and enzymes as being the prime causes of aging and carcinogenesis, phenomena that occur in eukaryotic organisms. He clearly presents the scientific evidence linking increased incidences of cancer in humans to increasing age.

Auerbach, Charles. *Mutation Research: Problems, Results, and Perspectives*. New York: John Wiley & Sons, 1976.
Auerbach's lengthy, thorough treatment of mutagenesis is one of the best introductions to the field. In chapter 1, "The History of Mutation Research," he describes important results of research demonstrating mutational effects such as crossing over, dosage effects, spontaneous mutations, and ultraviolet light damage to DNA. In chapter 3, "Classification of Gene Mutations," he discusses the various types of DNA damage and change. In chapters 4, "Reverse Mutations: Suppressors," and 8, "Repair of X-Ray Damage I. Repair of Chromosome Breaks," Auerbach addresses mechanisms by which cells may recover from mutational effects.

Brusick, David. *Principles of Genetic Toxicology*. New York: Plenum Press, 1980.
Brusick's work is an informative study of genetic mutation, the mechanisms of gene regulation within living organisms, and the means by which certain chemicals and radiations mutate genes, thereby affecting an organism and possibly its descendants as well. He examines the structure of DNA, the properties of various mutagenic substances, and mutations within both somatic and reproductive cells of organisms. He critiques the reliability of diagnostic screening tests to identify potential mutagens and provides twenty-one protocol tests.

Burck, K. B., E. T. Liu, and J. W. Larrick. *Oncogenes: An Introduction to the Concept of Cancer Genes*. New York: Springer-Verlag, 1988.
Burck, Liu, and Larrick discuss the mechanisms by which cancer arises in eukaryotic organisms, mechanisms that principally arise from genetic errors and/or rearrangements, events that can contribute equally to the further evolution

of a species. The process of cancer is a phenomenon that occurs in multicellular eukaryotes. It usually occurs, however, as a result of mutations within DNA brought about by viral DNAs, gene rearrangements, or by physical/chemical influences.

Butterworth, Byron E. *Strategies for Short-Term Testing for Mutagens/Carcinogens*. West Palm Beach, Fla.: CRC Press, 1979.
While mutation is a driving force in evolutionary change, scientists wish to understand the mechanisms of mutation in order to reduce mutational effects upon humans. Thirty scientists have contributed to this volume, which is a transcript of a 1979 Chemical Industry Institute of Toxicology workshop. Contributing authors include Byron Butterworth, Donald Clive, David Couch, Nancy L. Forbes, J. Patrick O'Neill, William Sheridan, and Gary Williams. The contributed papers include a critical review of experimental tests that attempt to identify mutagenic agents in the environment.

Celis, J. E., and J. D. Smith, eds. *Nonsense Mutations and tRNA Suppressors*. New York: Academic Press, 1979.
This work is a compilation of scientific papers that were presented at a July, 1978, Aarhus University symposium on biological mutations sponsored by the European Molecular Biology Organization. The papers deal primarily with the occurrence of a type of mutation in DNA called a nonsense, or missense, mutation. Such a mutation can be counteracted within the cell later on in the information flow process, specifically during the translation of protein from messenger RNA via an assistant molecule called transfer RNA (tRNA); mutant tRNA molecules can suppress the original mutation by inserting an appropriate amino acid into the gene-encoded protein. Contributing authors who explain this process include Sidney Altman, B. F. C. Clark, J. E. Celis, and J. D. Smith ("Suppressor tRNAs in Prokaryotes").

Collins, Andrew, Robert T. Johnson, and John M. Boyle, eds. *Molecular Biology of DNA Repair*. *Journal of Cell Science*, supp. 6. Cambridge, England: The Company of Biologists, 1987.
Twenty-two scientific papers presented at the 1986 DNA Repair Information Network and British Society for Photobiology in Manchester, England, constitute this volume. Topics include mutational damage to DNA, mutation and evolution, cellular DNA repair mechanisms, irradiation of cells and their DNA, and stress-induced changes in DNA. Contributors include Errol C. Friedberg, James B. Boyd, Michael H. L. Green, Alan R. Lehmann, Susan M. Povey, and Stuart Linn.

Committee on Chemical Environmental Mutagens, Board on Toxicology and Environmental Health Hazards, Commission on Life Sciences, National Research

Council (James F. Crow, chair). *Identifying and Estimating the Genetic Impact of Chemical Mutagens*. Washington, D.C.: National Academy Press, 1983.
This three-hundred-page report summarizes a 1979 to 1983 study of the effects of chemical mutagens upon human health. While focusing upon the effects of mutagens upon human somatic and reproductive cells, the results of the study are applicable to any species. The work describes the structure of DNA, the process of mutation, various identified chemical mutagens, experimental tests aimed at identifying possible chemical mutagens, and critiques of these tests. The committee concludes that short-term testing of mutagens is excellent; however, no methods for detecting human reproductive cell mutations currently exist.

Cooper, Geoffrey. *Oncogenes*. Boston: Jones & Bartlett, 1990.
Cooper, a leading molecular biologist and authority on the mechanisms of carcinogenesis, describes oncogenes, genes within cells that suddenly may alter and cause the cell to follow an unstable, cancerous course. He discusses the role of mutation in this process, as well as the contributing roles of certain viruses that infect cells. Oncogenes represent a class of genes whose role within cells and species evolution is poorly understood at present.

Friedberg, Errol C., and Philip C. Hanuaralt, eds. *Mechanisms and Consequences of DNA Damage Processing*. New York: Alan R. Liss, 1988.
This six-hundred-page work is a compilation of papers presented at a January, 1988, UCLA Symposium on Mutation held in Taos, New Mexico. Friedberg, Hanawalt, Susan S. Wallace, and Kenneth H. Kraemer head a list of more than four hundred participants who present papers and discuss topics relating to DNA, mutation, and cellular mechanisms for repairing mutational damage to DNA. While mutation is a driving force in cellular and species evolution, it often produces deleterious effects. Fortunately, the cell has evolved, by mutation, defense mechanisms for reducing the deleterious effects of mutational events from physical and chemical sources.

Generoso, W. M., M. D. Shelby, and F. J. de Gerres, eds. *DNA Repair and Mutagenesis in Eukaryotes*. New York: Plenum Press, 1980.
Sixty-four international researchers in the effects of mutagens on DNA met in Atlanta in June, 1979, at a conference sponsored by the National Institute of Environmental Health Sciences. Thirty-one summary and research papers are presented in this volume. They deal with the occurrence of mutations in eukaryotic species and with cellular mechanisms for negating the effects of mutations. R. F. Kimball, B. Strauss, and R. B. Getlow present reviews of the field. Alice Schroeder, Frederick J. de Serres, and others discuss mutational events in the pink bread mold *Neurospora crassa*. Bruce Baker, James Boyd, and others describe mutation in the fruit fly *Drosophila melanogaster*. Michael Bender, Koki Sato, and Miria Stefanini examine somatic cell mutations in mammalian species.

Hanauralt, Philip C., Errol C. Friedberg, and C. Fred Fox, eds. *DNA Repair Mechanisms*. New York: Academic Press, 1978.
An ICN-UCLA symposium on molecular and cellular biology held at Keystone, California, in 1978 produced this eight-hundred-page collection of 147 research papers. The detailed papers describe the processes of mutation in DNA, their evolutionary significance, and cellular defense mechanisms against mutagens and their effects. Peter A. Cerutti summarizes "Repairable Damage in DNA." Betsy M. Sutherland describes in detail "Enzymatic Photoreactivation of DNA." Bruce Ames, a leader in the field of mutation research, presents "Environmental Chemicals Causing Cancer and Genetic Birth Defects." Other contributors include Alan R. Lehmann, James B. Boyd, Franklin Hutchinson, B. A. Bridges, Ann Ganesan, and Raymond Waters.

Heddle, John A., ed. *Mutagenicity: New Horizons in Genetic Toxicology*. New York: Academic Press, 1982.
Heddle's work is an informative collection of scientific papers dealing with mutation and its effects upon organisms. Several leading research scientists in the field of mutation research and genetics have contributed to this volume. John Ashby addresses problems in mutation research in essay one, "Screening Chemicals for Mutagenicity: Practices and Pitfalls." Takashi Sugimura and Minako Nagao describe "The Use of Mutagenicity to Evaluate Carcinogenic Hazards in Our Daily Lives" in essay three. In essay eight, Veronica Maher and J. Justin McCormick discuss the "Measurement of Mutations in Somatic Cells in Culture."

Hedrick, Philip W. *Genetics of Populations*. Boston: Science Books International, 1983.
Hedrick's book is a good introduction to mutation and natural selection for or against mutated genes within populations of organisms. In chapter 1, "The Diversity of Genetic Variation," he discusses electrophoretic variants of certain enzymes (that is, genes) from various species, mutant alleles of genes, and lethal genes. In chapter 9, "Mutation," he addresses the causes and effects of mutations with individual genes and groups of genes. He relates "Population Genetics and Evolutionary Biology" in chapter 13.

Kilbey, B. J., M. Legator, W. Nichols, and C. Ramel, eds. *Handbook of Mutagenicity Test Procedures*. New York: Elsevier Scientific, 1977.
This excellent and essential source book for experimental procedures in assessing the mutagenicity of a particular unknown substance contains twenty-eight essays written by mutation research experts. Cited experiments involve a variety of case studies of mutagenic effects on various species, including humans. One classic paper in this book, "Methods for Detecting Carcinogens and Mutagens with the Salmonella/Mammalian Microsome Mutagenicity Test," is written by Bruce

Ames, Joyce McCorm, and Edith Yamasaki. Other contributors to this important volume include James E. Cleaver, M. S. Legator, F. K. Zimmermann, and H. B. Newcombe.

Lewontin, Richard C. *The Genetic Basis of Evolutionary Change*. Columbia Biological Series XXV. New York: Columbia University Press, 1974.
Lewontin, a prominent evolutionary geneticist, presents in this informative book a clear, not-too-complicated analysis of mutation, genetic variation, and natural selection. Lewontin describes difficulties in evolutionary genetics in chapter 2, "The Struggle to Measure Variation." He describes "The Genetics of Species Formation" in chapter 4. He stresses "The Genome as the Unit of Selection" in chapter 6.

Ravin, Arnold W. *The Evolution of Genetics*. New York: Academic Press, 1965.
This fine book for the beginning student is a gentle introduction to evolution and genetics. Ravin describes the evolutionary and mutational events that have led to the evolution of various species. In chapter 2, "The Use of Microorganisms in Molecular Genetics," he describes basic processes in bacterial life-styles such as conjugation, transduction, and transformation. In chapter 4, "Mutation and Recombination of Genetic Material," he addresses various chemical mutagens and their effects upon cellular DNA.

Rich, Marvin A., and Philip Furmanski, eds. *Biological Carcinogenesis*. New York: Marcel Dekker, 1982.
Robert Gallo, Carol A. Jones, Eliezer Huberman, and George A. Todaro head a list of prominent biologists who discuss the mechanisms of mutation and carcinogenesis in the papers presented in this important reference volume. The physical and chemical natures of mutagens and carcinogens are presented. Major emphasis is placed upon biochemical events in the cell, specifically within DNA, that contribute to mutant and cancerous cells.

Shepard, Thomas H. *Catalog of Teratogenic Agents*. 1973. 6th ed. Baltimore: The Johns Hopkins University Press, 1989.
Shepard, a physician, has assembled and described the mutagenic capabilities of 1,968 chemicals that are teratogenic, chemicals that produce birth defects when ingested by pregnant females. Each potential teratogen is characterized. Effects of various teratogenic substances upon various species (such as humans, chicks, or mice) are described during successive embryonic developmental stages. The teratogens range from derivatives of gasoline to locoweed to serotonin.

Sorsa, Marja, and Harri Vainio, eds. *Mutagens in Our Environment*. New York: Alan R. Liss, 1982.
Thirty-nine scientific essays on environmental mutagenic substances are presented

in this book, a compendium of papers presented at the twelfth annual meeting of the European Environmental Mutagen Society in Espoo, Finland, in 1982. Bruce Ames discusses "Carcinogens and Anti-Carcinogens." Ilse-Dore Adler describes "Germ-Cell Sensitivity in Mammals." G. M. Alink analyzes "Genotoxins in Waters." Gunter Obe stresses the "Mutagenicity of Our Life-Style."

Spiess, Eliot B. *Genes in Populations*. New York: John Wiley & Sons, 1977.
Spiess, a University of Illinois-Chicago Circle geneticist, has manufactured a lengthy book describing the processes of mutation within populations of organisms ranging from bacteria to humans. In chapter 1, he illustrates the "Genetic Network: The Castle-Hardy-Weinberg Conditions—One Pair of Alleles without Dominance." In chapter 13, "Mutations: Source of New Variation," he analyzes the role of DNA changes in the course of species evolution.

Walcher, Durain N., Norman Kretchmer, and Henry L. Barnett, eds. *Mutations: Biology and Society*. New York: Masson, 1978.
This scientific, but easy-to-follow, collection of thirty-one papers originally was presented at an International Organization for the Study of Human Development conference. Mutations, enzyme variations from species to species, gene variation and regulation, and genetic engineering are among the many important mutation related topics. Contributing scientists include Giorgio Bernardi, Raymond Boudon, Michele Carlier, Luigi L. Cavalli-Sforza, Evelyne Dumont, Roger Eckert, Rolf Ziegler, and Georges Larrouy.

Wallace, Bruce. *Basic Population Genetics*. New York: Columbia University Press, 1981.
Wallace, a population geneticist at Virginia Polytechnic and State University, discusses various aspects of genetics, mutation, and selection for and against various mutations within populations of organisms in this introductory textbook. He describes complex topics with great clarity and utilizes scientific data from the research literature. Graphs and mathematical models are found throughout the text to support key modes of thought.

Wallace, Bruce. *Fifty Years of Genetic Load: An Odyssey*. Ithaca, N.Y.: Cornell University Press, 1991.
The eminent population geneticist Wallace presents a scientific and philosophical presentation of genetics, mutation, and evolution in this intriguing book. In chapter 2, "Genetic Variation, Darwinian Fitness, and Genetic Load," he describes evolution using the simple mathematical Castle-Hardy-Weinberg equations. He discusses mutations and their effects in detail in chapters 3, "Studies of Irradiated Populations," and 4, "Random Mutations and Viability."

Chapter 6
THE PROKARYOTES

Bacteria

Beachey, E. H., ed. *Bacterial Adherence*. New York: Chapman & Hall, 1980.
Beachey's collection of scientific papers describes the abilities of various types of bacteria to adhere, or stick, to various surfaces, a phenomenon that enables these single-celled microorganisms to reproduce and to be transmitted from place to place via a number of host species. For example, Albert Collier describes the adherence of pneumococci to the slippery mucous membranes of the human respiratory tract. Shigeyuki Hamada and Hutton D. Slade examine the ability of *Streptococcus mutans* to stick to smooth surfaces within the bodies of host animals.

Bonner, John Tyler. *Morphogenesis: An Essay on Development*. New York: Atheneum, 1963.
This interesting work by a leading evolutionary scientist elaborates upon the physical dimensions and properties of cell structure and function, beginning with simple prokaryotic cells such as bacteria and then advancing to more complicated life forms. Chapter 2, "Size and Pattern," examines the fact that volume increases as a cube of linear dimensions, whereas area increases as a square, thereby limiting the maximum size of cells. Chapters 4, "Patterns of Growth," and 5, "Patterns of Morphogenetic Movements," compare cellular growth and movement to crystalline growth.

Bousfield, I. J., and A. G. Callely, eds. *Coryneform Bacteria*. New York: Academic Press for the Society for General Microbiology, 1978.
J. M. Hardie, M. D. Collins, and D. E. Minnikin provide descriptive research survey articles in this volume devoted to one class of pathogenic bacteria: the coryneforms. The survey articles describe the characteristics of coryneform bacteria, the diseases for which they are responsible, molecular studies of their genetic machinery and structure, their cell wall and lipid composition, and numerical taxonomy of coryneform species. A few papers are devoted to the computer identification of specific coryneform bacterial types.

Brieger, E. M. *Structure and Ultrastructure of Microorganisms: An Introduction to a Comparative Substructural Anatomy of Cellular Organization*. New York: Academic Press, 1963.
Brieger's book is a clear description of the physical structure and functioning of the components of the bacterial cell. He starts with the "Ultrastructure of the Cell" in chapter 1. Chapters 2, "The Nuclear Envelope," and 3, "Chromatin and

Chromosomes," deal with the genetics of bacterial cells. The cytoplasmic membrane and cell walls of bacteria are discussed in detail in chapters 24, "The Cytoplasmic Membrane in Bacteria," and 25, "Fine Structure of the Cell Wall." Brieger also examines the evolution of mitochondria and plastids in eukaryotic cells.

Broda, Paul. *Plasmids*. San Francisco: W. H. Freeman, 1979.
Broda's authoritative work on plasmids is a valuable reference for students of extrachromosomal DNA. Plasmids are circularized, small DNA molecules that contain several dozen genes and that occupy the cytoplasms of both prokaryotes and eukaryotes. Broda describes the discovery of plasmids in bacteria, the isolation of plasmids by density gradient centrifugation, transformation and transduction experiments with plasmids, and the applications of plasmids to cloning genes. He also describes the evolution of plasmids within living cells and the role of plasmid genes in conferring antibiotic resistance to certain strains of pathogenic bacteria, thereby affecting certain human diseases.

Chambliss, Glenn, and James C. Vary, eds. *Spores VII*. Washington, D.C.: American Society of Microbiology, 1978.
This book is a collection of very scientific papers describing the molecular biology of sporulating bacteria, bacteria such as *Bacillus subtilis*, which can produce a protective spore coat around their cells during periods of environmental stress and thereby lie dormant almost indefinitely until conditions improve. These papers emanate from the seventh International Spore Conference held in Madison, Wisconsin, in October, 1977. Contributing scientists include Alex Keynan, James Vary, Kenneth Bott, Pranab K. Mukherjee, and Maureen A. Shiflett.

Cooper, Stephen. *Bacterial Growth and Division: Biochemistry and Regulation of Prokaryotic and Eukaryotic Division Cycles*. San Diego: Academic Press, 1991.
In this excellent five-hundred-page work, Cooper introduces the molecular biology of bacteria with great clarity. He describes essential cellular processes such as DNA replication, the binary fission of a cell to form two progeny cells following DNA replication, cell surface synthesis by bacteria, and models of bacterial growth. Cooper investigates the genetic regulation of bacterial cell processes as well.

Doelle, H. W. *Bacterial Metabolism*. New York: Academic Press, 1969.
Doelle's extensive study of the biochemistry and metabolic activities of various bacterial species is an excellent introduction to the field. Doelle examines essential processes that occur within the cells of various bacterial species, in the process often comparing and contrasting each species considered. For example, aerobic (oxygen-consuming) bacteria are metabolically compared to anaerobic (non-oxygen consuming) bacteria. Other aspects of bacterial biochemistry that are examined

include photosynthesis in the photoautotrophic bacteria, carbohydrate metabolism, fermentation, nitrogen metabolism in nitrogen fixators, and the utilization of metals by various bacterial species.

Dubnau, David A., ed. *Bacillus subtilis*. Vol. 1 in *The Molecular Biology of the Bacilli*. New York: Academic Press, 1982.
This collection of interesting scientific essays and discussions focuses upon the sporulation bacterium *Bacillus subtilis*, a highly successful and generally harmless species to humans. Genetic control mechanisms, specifically with respect to how the bacterium produces its spore coat in response to a hostile environment, are studied in detail in the papers. Contributing scientists include Roy H. Doi, Issar Smith, David Dubnau, and Richard Losick. Losick examines "Sporulation Genes and their Regulation."

Dubourguier, H. C., et al., eds. *Biology of Anaerobic Bacteria*. Progress in Biotechnology 2. New York: Elsevier, 1986.
This book is a collection of detailed scientific papers analyzing the anaerobic bacteria, prokaryotes that do not metabolize molecular oxygen and that very probably are representative of the earliest life forms on earth more than three billion years ago. The papers included in this volume were presented at a June, 1986, International Seminar on the Biology of Anaerobic Bacteria in Lille, France. Molecular and biochemical analyses of anaerobic bacteria are presented by microbiologists such as A. Boulahrouf, G. Fonty, M. B. Romond, and C. Romond.

Dworkin, Martin. *Developmental Biology of the Bacterium*. Reading, Mass.: Benjamin/Cummings, 1985.
Dworkin, a University of Minnesota professor, examines various aspects of bacterial cell development and the genetic control of developmental processes in this excellent short work on the subject. In chapter 9, "The Molecular Basis of Morphogenesis," he specifically looks at gene control systems that may trigger and/or regulate changes in bacterial metabolism, reproduction, and so forth. He continues this study in chapter 10, "Regulation of Development." Great photographs and descriptive models of various bacterial metabolic processes are provided.

Fraenkel-Conrat, Heinz, ed. *Molecular Basis of Virology*. American Chemical Society Monograph 164. New York: Reinhold, 1968.
In this lengthy, scientific work, Fraenkel-Conrat and numerous other microbiologists study the molecular structure of viruses, their origin and evolution, possible origins within prokaryotic cells, and the various types of viruses, including DNA and RNA phage. Contributing scientists to this volume include Fraenkel-Conrat, Lloyd M. Kozloff, Anne Joyner, E. R. Pfefferkorn, and Aaron J. Shotkin. The

biosynthesis of viruses within host cells is covered in great detail in several important papers.

Fuller, R., and D. W. Lovelock, eds. *Microbial Ultrastructure: The Use of the Electron Microscope*. New York: Academic Press, 1976.
The Society for Applied Bacteriology Autumn, 1974, meeting launched this collection of papers that present the detailed cellular and subcellular structures of bacterial cells in great detail and beauty. The texts of the various articles are greatly enhanced by transmission electron microphotographs that magnify the bacteria up to hundreds of thousands of times. The applications of electron microscopy to understanding bacteria is clearly presented in papers dealing with such varied topics as the sex pili of certain bacterial species and the accumulation of dental plaque by successive layers of bacterial cells.

Glass, Robert E. *Gene Function: E. Coli and Its Heritable Elements*. Berkeley: University of California Press, 1982.
In this impressive work, Glass describes all aspects of the human intestinal bacterium *Escherichia coli*. He describes the structures of this bacterial cell, its growth, its symbiotic relationship with its host, its nutrient requirements, gene regulatory mechanisms (such as the lactose and tryptophan operons), and the flow of information through the cell from DNA to RNA to protein. He describes mutational events in the bacterial genes and their effects upon the cell. He also addresses plasmid DNA and bacteriophage (virus) infections of the bacterial cells.

Gregory, Philip H. *The Microbiology of the Atmosphere*. New York: John Wiley & Sons, 1973.
This interesting scientific work, very understandable for the layperson, is a unique view of microorganisms as atmospheric organisms that can drift airborne for indefinite periods of time on specks of dust, aerosols, and the like. In chapter 2, he describes bacterial "Spores: Their Properties and Sedimentation in Still Air." He elaborates upon the extensive abundance of microorganisms in air in chapters 3, "The Atmosphere as an Environment," and 5, "The Air Spora Near the Earth's Surface."

Ierusalimskii, N. D. *Fundamentals of the Physiology of Microorganisms*. Israel: Israel Program for Scientific Translations, 1966.
Available from the U.S. Department of Commerce Clearinghouse for Federal Scientific and Technical Information. In this detailed scientific work, the Russian microbiologist Ierusalimskii examines the biochemistry of prokaryotic cells, their metabolism, mechanisms of DNA replication, binary fission of prokaryotic cells, RNA transcription, protein translation, and the effects of environmental changes upon prokaryotic cells. The work is comprehensive, but it is written and translated with great clarity for the student of biology.

Ingraham, John L., Ole Maaloe, and Frederick C. Neidhardt. *Growth of the Bacterial Cell.* Sunderland, Mass.: Sinauer Associates, 1983.
This outstanding introduction of microbial cell biology and genetics, written by three pioneers in the field of molecular genetics, describes the complicated, yet still poorly understood, biochemistry of bacterial cells. The authors describe molecular genetic control mechanisms within the genes of bacterial chromosomes, including a good discussion of the multiple-gene operons (such as lactose operon and tryptophan operon). Each chapter contains excellent diagrams and chapter summaries.

INSERM—Societe Francaise de Microbiologie. *Spheroplasts, Protoplasts, and L-Forms of Bacteria.* Foreword by Jacques Roux. Paris: Author, 1977.
Based upon a September, 1976, colloquium in Montpellier, France, the contributed papers in this volume deal with novel forms and structures of bacteria, specifically those species that develop exotic membranes and walls that surround and protect the cell as well as contribute to nutrient assimilation. The scientific papers are presented in English and contain excellent microphotographs of bacterial spheroid formations. Contributors include Sarabelle Madoff, W. W. Gregory, Harry Gooder, Jacques Roux, Priscilla B. Wyrick, and John W. Lawson.

Keilin, David. *The History of Cell Respiration and Cytochrome.* Prepared by Tom Keilin. New York: Cambridge University Press, 1966.
The late scientist Keilin describes the process of cellular respiration, a series of chemical reactions that are essential for the oxidative production of the energy molecule adenosine triphosphate from chemical information found within sugar. Keilin describes the history of respiration research, beginning with the ancient Greeks and progressing to modern knowledge. He describes clearly the basic processes of cellular respiration in prokaryotes and eukaryotes, including details of the molecules cytochrome and eukaryotic hemoglobin.

Kihlman, Bengt A. *Actions of Chemicals on Dividing Cells.* Englewood Cliffs, N.J.: Prentice-Hall, 1966.
This short, detailed work summarizes the basic chemical properties and functions of the cell and the effects of various chemicals upon the occurrence of these chemical reactions. Chapter 2 describes "The Chemistry and Physiology of the Interphase Nucleus," dealing with DNA. Chapter 5, "DNA Synthesis at the Cytological Level and Chromosome Duplication," discusses the control of cellular activities by the DNA molecule and how chromosome rearrangements can affect such control. Chapter 14, "DNA, Mutation, and Chromosomal Aberrations," looks at the effects of chemicals and radiation upon the functioning and evolution of the cell.

Lehninger, Albert L. *Bioenergetics: The Molecular Basis of Biological Energy Transformations*. New York: W. A. Benjamin, 1965.
Lehninger's introduction to basic cellular energy biochemistry is an excellent survey of the subject for the beginning student. Chapter 2, "Principles of Equilibrium Thermodynamics," explores the cellular control of energy metabolism. Chapter 3, "The Flow of Energy in the Biological World," demonstrates the interconnectedness between all organisms on earth in the cycling of energy through the environment. Chapter 6, "Respiration and ATP Formation in the Mitochondrion," shows how oxygen and food are used to manufacture cellular energy in eukaryotes, a phenomenon that first evolved in the prokaryotes. The book contains excellent microphotographs.

Levinson, Hillel S., A. L. Sonenshein, and Donald J. Tipper, eds. *Sporulation and Germination*. Washington, D.C.: American Society for Microbiology, 1981.
Ernst Freese, James C. Garbe, Patricia S. Vary, R. C. Goldman, Ellen Kraig, and Susanne L. Simon are among the dozens of scientists who contributed research papers to this volume based upon proceedings at the eighth International Spore Conference at Woods Hole, Massachusetts, in October, 1980. The articles deal with bacterial and fungal species that can produce spores capable of withstanding extreme drought and lack of oxygen for long periods of time; sporulation enables such species to withstand long periods of environmental stress. The biochemical mechanisms of sporulation for species such as *Bacillus subtilis* and the pink bread mold *Neurospora crassa* are explored in these articles.

Losick, Richard, and Lucy Shapiro, eds. *Microbial Development*. Cold Spring Harbor, N.Y.: Cold Spring Harbor Laboratory, 1984.
Several hundred microbiologists attended this 1983 Conference on Microbial Development at the Cold Spring Harbor Biological Laboratory directed by Nobel laureate James D. Watson. Major topics discussed in the presentation papers include bacterial cellular gene regulation, cell differentiation in primitive eukaryotes, the roles of pheromones (extraorganismal hormones) in gene control, multicellular development, and pattern formation in species such as the cellular slime mold *Dictyostelium discoideum*. Contributors include H. K. MacWilliams, C. N. David, Brent Ely, William D. Donachie, Kenneth J. Begg, and Neil F. Sullivan.

Mandelstam, Joel, and K. McQuillen, eds. *Biochemistry of Bacterial Growth*. New York: John Wiley & Sons/Blackwell Scientific Publications, 1968.
Mandelstam and McQuillen are among the many contributors of scientific papers to this volume describing bacterial biochemistry, control of gene regulation in bacteria, bacterial metabolism and physiology, and cellular division. Classification schemes for bacterial species and the differentiation of bacterial populations are discussed. Other contributors to the volume include Michael J. Waring, Edward Dawes, and J. R. S. Fincham.

Meynell, G. G. *Bacterial Plasmids: Conjugation, Colicinogeny, and Transmissible Drug-Resistance*. Cambridge, Mass.: MIT Press, 1973.
This short book (164 pages) incorporates the results of 672 references to describe the extrachromosomal plasmids that contain certain essential genes for bacterial metabolism, but which are located in the cellular cytoplasm, not with the chromosomal genes. These plasmids include F and R types that enable the bacterium to mate with non-F or non-R bacteria. Some of these plasmids contain mutant genes whose protein byproducts confer upon the cell resistance against antibiotics, an area of much concern for the medical community.

Minton, Nigel P., and David J. Clarke, eds. *Clostridia*. Biotechnology Handbook 3. New York: Plenum Press, 1989.
Jan R. Andreesen, Elizabeth P. Coto, Erko Stackebrandt, Clifford C. Shone, Peter Hambleton, Raphael Lamed, Juergen Wiegel, and numerous other scientists contribute to this volume concerning the clostridial bacteria, a group of bacterial species that release deadly toxins within their hosts during infections. To construct evolutionary phylogenies, the papers examine the genetic regulation of toxin production as well as comparisons of toxins between various species.

Mishustin, E. N., and V. K. Shil'nikova. *Biological Fixation of Atmospheric Nitrogen*. Translated by Alan Crozy. University Park: Pennsylvania State University Press, 1971.
In this thorough but clear volume, two Russian biochemical microbiologists describe the biology of the nitrogen-fixing bacteria, bacterial species that convert atmospheric molecular nitrogen into nitrogen compounds useful for plant growth. As such, the authors emphasize the symbiotic relationship between these bacterial species and certain plant species (such as peanuts). They examine the biochemistry of nitrogen fixation in bacterial species such as azotobacter, beijerinckia, and clostridia, the last being a facultative anaerobe.

Moat, Albert G. *Microbial Physiology*. New York: John Wiley & Sons, 1979.
Moat begins this lengthy, but clear, introduction to prokaryotic biochemistry with an informative chapter 1, "Classification of Microorganisms." He then proceeds to bacterial and protistan metabolisms in chapters 2, "Cell Wall," and 3, "Anaerobic Pathways." Moat approaches the transfer of cellular information from DNA to RNA to protein in later chapters; he presents these concepts with tremendous clarity. Chapter 9, "Effect of Toxic Environment," describes situations that inhibit the proliferation of microorganismal species.

Nester, Eugene W., et al. *Microbiology*. Philadelphia: Saunders College Publishing, 1983.
This book is a good introduction to microorganisms for the layperson. It covers all types of microorganisms ranging from viruses to bacteria to protozoa to fungi.

Chapter 3 describes the "Functional Anatomy of Prokaryotic and Eukaryotic Cells." Chapter 12 presents "Bacteria" in detail, including those species that are pathogenic (disease-producing). The authors address "Terrestrial Microbiology" and "Aquatic Microbiology" in chapters 28 and 29, respectively.

Perez-Miravete, A., coordinator. *Behaviour of Micro-Organisms*. New York: Plenum Press, 1983.
The detailed scientific papers in this book emerged from the tenth International Conference of Microbiology, held in Mexico City in August, 1970. The interactions of various microorganisms, including bacteria, fungi, and protozoa, with each other, with animals and plants, and with changing environmental conditions are extensively cataloged. The biochemical and metabolic bases of microbial behavior is emphasized throughout the many papers. Contributing scientists include D. D. Woodward, J. W. Hastings, and E. M. Eisenstein.

Sistrom, W. R. *Microbial Life*. New York: Holt, Rinehart and Winston, 1962, 1969.
In this valuable reference, Sistrom presents the world of microorganisms with special reference to the prokaryotic bacteria and the eukaryotic protists. Chapter 3, "The Bacterial Cell," demonstrates that bacteria, while being single-celled, are far more complicated than they look. Chapter 6, "Growth of Bacteria and the Synthesis of Macromolecules," details the biochemistry of the bacterial cell in simple terms. Chapter 9, "Bacteria and the Economy of Nature," shows the importance of bacteria within the environment. Excellent flowcharts and diagrams supplement the short, concise chapters.

Sokatch, J. R. *Bacterial Physiology and Metabolism*. New York: Academic Press, 1969.
Sokatch provides an elaborate, but lucid, coverage of the biochemistry behind the inner cellular workings of prokaryotes. Sokatch presents a very understandable study of bacterial genetics and gene control mechanisms, complicated enzyme-mediated biochemical pathways (such as glycolysis), electron transport mechanisms in adenosine triphosphate synthesis, and nutritional requirements of bacteria. This book is an excellent sourcebook for the serious biology student.

Stephenson, Marjory. *Bacterial Metabolism*. 1949. 3d ed. Cambridge, Mass.: MIT Press, 1966.
Stephenson's extensively referenced text provides an authoritative presentation of the essential biochemical and metabolic processes that keep bacterial cells alive and proliferating and that have played key roles in the evolution of all living forms on the earth. Stephenson addresses topics such as the growth and nutritional requirements of bacteria, the metabolic synthesis and recycling of nucleic acids (DNA and RNA), cellular respiration to produce usable energy from sugar, and anaerobic fermentation reactions during decomposition of formerly living material.

She describes with clarity the evolution of autotrophic (photosynthetic) organisms more than two-and-one-half billion years ago.

Thimann, Kenneth V. *The Life of Bacteria: Their Growth, Metabolism, and Relationships*. 1955. 2d ed. New York: Macmillan, 1963.
Thimann's exhaustive coverage of bacteria describes their biochemistry, metabolism, and physiology in exceptional depth. The nine-hundred-page work is extensively referenced, and it provides a clear introduction to the subject for the advanced microbiology and biochemistry student. Thimann elaborates upon topics such as bacterial cellular respiration, both aerobic and anaerobic, the evolution of autotrophs, the metabolism of soil bacterial species, and the laboratory culturing of bacteria for research purposes.

Umbreit, W. W. (Rutgers University), and the Class of 1975 (Douglass College). *Essentials of Bacterial Physiology*. Stroudsburg, Pa.: Dowden, Hutchinson & Ross, 1976.
This tremendous introductory textbook, written by a microbial biochemist and his students, describes the physiological constraints that influence bacterial growth and proliferation. Umbreit provides an extensive supply of data and numerous models to illustrate key points. Bacterial energy metabolism via anaerobic and aerobic cellular respiration reactions, environmental influences upon bacteria, development of bacterial cell walls, and identification procedures for classifying bacteria (such as the gram stain test) are among the many topics covered.

Van Iterson, Woutera, ed. *Inner Structures of Bacteria*. New York: Van Nostrand Reinhold, 1984.
Fifty-nine research papers highlight this important volume, which explores the submicroscopic anatomy of bacterial cells. Fantastic microphotographs and transmission electron micrographs provide compelling support to the text of most of the papers. The articles are elaborate in content, although many are comprehensible for the interested student. Contributing authors to the volume include van Iterson, C. F. Robinow, D. J. Mason, C. C. Rensen, S. W. Watson, and A. J. Verkleiz.

Weinrich, D. H., Ivey F. Lewis, and John R. Raper, eds. *Sex in Microorganisms*. Washington, D.C.: American Association for the Advancement of Science, 1954.
This excellent, clear discussion of the then-recently discovered sexual cycle in some bacteria contains position papers by several prominent molecular biologists. Sex in bacteria occurs when certain bacterial types containing a specific plasmid transfer genetic information to other bacterial cells that lack this particular plasmid. The phenomenon was elucidated first by Joshua Lederberg and Edward Tatum, both eventual Nobel laureates who have papers in this volume. Other contributors include Ruth Patrick, Ralph A. Levin, and David Nanney.

Wistreich, George A., and Max D. Lechtman. *Microbiology*. New York: Macmillan, 1988.

Wistreich and Lechtman's *Microbiology* is perhaps the most detailed, comprehensive introduction to the subject available. They provide in-depth information on hundreds of bacterial, viral, fungal, protozoan, and helminth microorganisms, many diseases caused by some of these organisms, and diagnosis and treatment of microorganismal diseases. The elaborate text is supported by graphic photographs of disease microorganisms and their human hosts, summary viral and bacterial charts arranged by body organ system, disease challenge exercises, and clear descriptions of microbiological techniques. Bacterial anatomy and physiology are presented in chapter 7, "Prokaryotes: Their Structure and Organization." Bacteria and viruses that affect the respiratory, intestinal, cardiovascular, urinary, and central nervous systems are covered in independent chapters.

Wood, E. J. Ferguson. *Marine Microbial Ecology*. New York: Reinhold, 1965.

In this excellent work, Wood describes in detail with great clarity the distributions of bacteria, algae, plankton, and protozoa in the earth's oceans. He begins with a presentation of environmental conditions in "The Marine Environment," chapter 2. In chapter 3, "The Micro-Organisms," Wood describes the thousands of different microorganismal species that inhabit the oceans. Chapter 4, "Aspects of Productivity," examines the essential contributions of microorganisms to the earth's environment on a global scale.

Youvan, Douglas C., and Feirzi Daldal, eds. *Microbial Energy Transduction: Genetics, Structure, and Function of Membrane Proteins*. Cold Spring Harbor, N.Y.: Cold Spring Harbor Laboratory, 1986.

A February, 1986, bacterial metabolism conference at the prestigious Cold Spring Harbor Biological Laboratories on Long Island attracted hundreds of scientists from several countries. The papers presented in this volume describe the biochemistry and biophysics of proteins that lie in the bacterial cellular plasma membranes, proteins that mediate nutrient intake into the cell, communication with other cells, and general purpose functions. The encoding of these proteins by the bacterial genes and the control of these genes are elaborated as well. Contributing scientists include Charles Arntzen, Robert Huber, Samuel Kaplan, Gregory R. Moe, Melvin Simon, and Herbert Zuber.

The Evolution of Autotrophs

Battley, Edwin H. *Energetics of Microbial Growth*. New York: John Wiley & Sons, 1987.

Battley's interesting book is a historical approach to the scientific study of microbes. Battley elaborates upon the bacterial cell's nutritional and energy requirements as well as the biochemistry of the cellular chemical reactions that serve to acquire extracellular substances for the interconversion of energy. Battley cites his own work as well as the contributions of his colleagues (such as H. Tamiya). He compares the metabolisms of aerobic and anaerobic cellular respirations, both processes that evolved very early in the history of cellular life on earth.

Beevers, Harry. *Respiratory Metabolism in Plants*. New York: Harper & Row, 1961.
Beevers describes a very important series of chemical reactions in prokaryotic cells that provides energy and is essential for cell survival: the chemical reactions of glycolysis and the citric acid cycle, the latter also being termed the Krebs cycle and the Tri-Carboxylic Acid cycle. Beevers discusses "The Energetics of Respiration" (chapter 5) by illustrating the production of the energy molecule adenosine triphosphate (ATP) from the information provided by simple sugars. He describes with great precision these intricately detailed chemical reactions for plants, which are eukaryotes, although the reactions are equally as applicable to most prokaryotic cells, since prokaryotic cells evolutionarily were the first cells to develop these reactions of cellular respiration.

Calvin, Melvin, and J. A. Bassham. *The Photosynthesis of Carbon Compounds*. New York: W. A. Benjamin, 1962.
In this book, the eminent biochemists Calvin and Bassham describe the intricate chemical reactions of photosynthesis, a phenomenon that first developed in the blue-green photobacteria and that later was incorporated by single-celled eukaryotic plankton and multicellular eukaryotic plants. They discuss both the light and light-independent chemical reactions of photosynthesis, the latter series of reactions also being called the Calvin-Benson cycle, appropriately named after its discoverer (the author). Photosynthesis was a key event in the evolution of life on earth because it made the planet's life self-supporting.

Danks, Susan M., E. Hillary Evans, and Peter A. Whittaker. *Photosynthetic Systems: Structure, Function, and Assembly*. New York: John Wiley & Sons, 1983.
Danks, Evans, and Whittaker present a tremendous account of living organisms on earth, the evolution of photosynthesis, and the basic chemical reactions of the process in this short, concise work. Chapter 1, "Introduction," surveys both prokaryotic and eukaryotic cells, electron transfer in photosynthesis, and colored pigment molecules that absorb light in the photosynthetic process. Chapter 2, "Photosynthetic Phosphorylation," looks at different photosynthetic pigments (such as chlorophyll, phycobillins, and carotenoids). Chapter 4 details "The Origin and Assembly of Chloroplasts," an important step that led to the evolution of eukaryotic plants.

Dawes, Edwin A. *Microbial Energetics*. New York: Chapman & Hall, 1986.
Dawes's exciting book is a short, concise description of bacterial bioenergetics, the interconversion of energy for cellular chemical reactions through the recycling of adenosine triphosphate from adenosine diphosphate and inorganic phosphate. Dawes provides a simple introduction to Peter Mitchell's chemiosmotic theory of energy transduction across cell membranes. He takes complicated biochemical and biophysical phenomena and translates them into layperson's terms.

Foyer, Christine H. *Photosynthesis*. New York: John Wiley & Sons, 1984.
In this clear book containing many illustrations and electron microphotographs of chloroplasts in plants, Foyer describes the evolution of photosynthesis first in prokaryotes and then its elaboration in eukaryotic species such as plankton and plants. The various photosystems and chemical reactions are analyzed along with the light-absorbing pigment molecules. Alternative photosynthetic reactions are considered, including the four-carbon pathway used in certain plants (such as crabgrass or corn).

Gest, Howard, Anthony San Pietro, and Leo P. Vernon, eds. *Bacterial Photosynthesis*. Yellow Springs, Ohio: Antioch Press, 1963.
This collection of scientific papers was presented at a 1963 symposium held at the Charles F. Kettering Research Laboratory. The theoretical and research papers in this volume are very clear, do not contain excess scientific jargon, and convey the ideas of many brilliant biochemists and microbiologists concerning the mechanisms of photosynthesis, a process that evolved first in bacteria more than two billion years ago and that is essential for the continuation of all life on earth. Contributing scientific authors to this volume include Britton Chance, Mitsuo Nishimura, Martin D. Kamen, Leo P. Vernon, and Hans Gaffron.

Olson, J. M., et al., eds. *Green Photosynthetic Bacteria*. New York: Plenum Press, 1988.
An August, 1987, European Molecular Biology Organization (EMBO) workshop produced the survey and research papers in this informative reference volume. Papers are grouped subject by subject as they relate to various aspects of photosynthesis in bacteria. Part 6, "Phylogeny and Taxonomy," categorizes the various types of photosynthetic bacterial species, as elaborated by authors such as E. Stackebrandt. Part 4 deals with the essentials of "Electron Transport," part 7 discusses the "Physiology and Ecology" of photosynthetic bacteria, and part 8 analyzes the "Genetics" of such bacteria. Other contributors include H. W. J. Smit and H. Zuber.

Steinback, Katherine E., Susan Bonitz, Charles J. Arntzen, and Lawrence Bogorad, eds. *Molecular Biology of the Photosynthetic Apparatus*. Cold Spring Harbor, N.Y.: Cold Spring Harbor Laboratory, 1985.

This very molecular series of papers was presented at a NATO Advanced Research Workshop held at Cold Spring Harbor Biological Laboratory on Long Island, New York, in 1984. The detailed papers cover the chemistry of photosynthesis, photosystems I and II, cytochromes, photosynthetic pigment molecules, and ATP generation/use. Contributing authors include the editors, Bertil Andersson, David Kyle, Catherine P. Chia, Gunter Hauska, Elisabeth Gantt, and Jeffrey Palmer.

Chapter 7
INTRACELLULAR PARASITES: THE VIRUSES

Andrewes, Christopher H. *The Natural History of Viruses*. New York: W. W. Norton, 1967.
In this very simple, but thorough, work, Andrewes describes the history of virus research from the late 1800's to the present. He describes the lytic and lysogenic phases to the typical virus life cycle, the ecology of viruses and their host cells, and the vectors (transmitters) that spread viruses from host to host. He lists numerous species and describes the various diseases with which each virus is associated. He also describes oncogenic (cancer-causing) viruses. Excellent microphotographs of viruses support the clear text.

Andrewes, Christopher H., and H. G. Periera. *Viruses of Vertebrates*. Baltimore: Williams & Wilkins, 1964, 1967, 1972.
Andrewes and Periera's chronology of vertebrate viruses is a step-by-step analysis of different viral types that cause disease in vertebrate animals. Special attention is focused upon viruses that afflict humans and other primates. Part 1 summarizes "RNA Viruses," part 2 looks at "DNA Viruses," and part 3 examines uncharacterized viruses such as the hepatitis group viruses. This work is a valuable virus reference book.

Baltimore, David, Alice S. Huang, and C. Fred Fox, eds. *Animal Virology*. New York: Academic Press, 1976.
Baltimore, Huang, and Fox's work is a collection of important research papers by leading animal virologists concerning the many diverse viral types that parasitize the cells of animal species, both invertebrate and vertebrate. Extensive molecular biological data is found throughout this 824-page volume. Contributing authors include such eminent virologists as J. Michael Bishop, D. H. Gillespie, George Todaro, David Smotkin, and Robert Sampson.

Botchan, Michael, Terri Grodzicker, and Phillip A. Sharp, eds. *Cancer Cells 4-DNA Tumor Viruses: Control of Gene Expression and Replication*. Cold Spring Harbor, N.Y.: Cold Spring Harbor Laboratory, 1986.
This large, 620-page work represents the proceedings of a 1985 meeting concerning the genetics of DNA tumor viruses held at the Cold Spring Harbor Biological Laboratory on Long Island, New York. More than one hundred conference participants are represented in the papers included in the volume. DNA tumor viruses such as Simian Virus 40, the Epstein-Barr Virus (the cause of mononucleosis and Burkitt's lymphoma), and polyomavirus are discussed in detail. Viral processes such as RNA transcription, protein translation, and cellular transformation by infecting viruses are intricately described as well.

Brooks, Stewart M. *The World of the Viruses*. New York: A. S. Barnes, 1970.
In this simple introduction to viruses for the layperson, Brooks describes all aspects of the viral life cycle, the mechanisms by which viruses are transmitted (such as wind, water, direct contact, or insects), and the diseases caused by viruses. He describes the elementary biochemistry of viruses, their lytic mode of reproduction, and their dormant lysogenic phase. He discusses numerous species of viruses and their host cells.

Bukhari, A. I., J. A. Shapiro, and S. L. Adhya, eds. *DNA: Insertion Elements, Plasmids, and Episomes*. Cold Spring Harbor, N.Y.: Cold Spring Harbor Laboratory, 1977.
This extremely detailed, eight-hundred-page volume contains research and summary papers that were presented at a May, 1976, symposium on DNA insertion at the Cold Spring Harbor Laboratory. Contributors to the volume include the editors, P. Starlinger, N. D. F. Grindley, A. E. Jacob, D. E. Berg, M. Nomura, P. A. Peterson, and R. F. Baker. They discuss extrachromosomal DNA, small DNA molecules that exist in the cellular cytoplasm or that remaneuver themselves from one chromosomal location to another. Elaborate genetic maps and photographs of DNA from many different species are presented.

Calendar, Richard, ed. *The Bacteriophages*. Vols. 1-2. New York: Plenum Press, 1988.
In this comprehensive survey work, Calendar assembles contributed essays from leading molecular virologists that describe the many diverse types of bacteriophage, viruses that infect bacterial cells. The detailed articles describe many aspects of viral life cycles, including both DNA and RNA phage and lytic/lysogenic cycles. In volume 1, scientists such as Allan Campbell, Henry Drexler, Margarita Salas, and Huber R. Warner describe bacteriophage mu, T1, T5, T7, P1, SPO1, and archaebacteriophage. In volume 2, Yen Sen Ho, Fred Eiserling, Eugene Martin, John R. Murphy, and others discuss bacteriophage N4, P4, T4, and OX174.

Carney, Desmond, and Karol Sikora, eds. *Genes and Cancer*. New York: Wiley, 1990.
Carney and Sikora have assembled a detailed, yet clear, analysis of oncogenes, segments of DNA in eukaryotic cellular chromosomes and in some virus chromosomes that are capable of converting a normal, differentiated cell into a rapidly multiplying cancer. Karol Sikora, William Gullick, and Suzanne A. Eccles are among the many scientists who discuss oncogenes and anticancer therapies involving antibodies, drugs, and growth factors.

Ciba Foundation. *Novel Diarrhoea Viruses*. New York: John Wiley & Sons, 1987.
This report is a collection of papers presented at the 128th Ciba Symposium in

London in 1986. The detailed papers discusses the molecular biology of viruses that cause symptoms of diarrhea in their animal hosts, viruses such as adenovirus, various RNA viruses, and rotavirus. The infectious nature of these viruses upon their host's intestinal cells and medical treatments for these infections are described with great thoroughness. Contributing scientists include R. F. Bishop (chair), T. H. Flewett, H. B. Greenberg, and N. R. Blacklow. Each article includes commentaries, group discussions, and concluding remarks by chairperson Bishop.

Cooper, J. I. *Viruses and the Environment*. New York: Chapman & Hall, 1984.
Cooper's short work is an introduction to the science of virology, the study of viruses, their properties, life cycles, origins, and effects upon the cellular life that they parasitize. Viruses are noncellular entities that reproduce en masse within living cells. They are always parasites. Cooper examines their distribution and spread via air, water, land, and direct transfer between animals, etc. Various vectors (carriers) are described, including animals. Animal immunity to viruses is studied, and the spread of viruses to cells is analyzed using predator-prey models.

Crowell, Richard L., and Karl Lonberg-Holm, eds. *Virus Attachment and Entry into Cells*. Washington, D.C.: American Society for Microbiology, 1986.
The articles in this comprehensive work cover a very important aspect of viral infection of cells, the attachment of the virus particle to the cell's protective membrane. Different viruses attack different host animal, plant, fungal, and bacterial species by keying on specific molecules within the host cell membranes. The papers herein treat topics such as cellular membrane receptor proteins for viruses, attachment viral proteins, and mechanisms of infection following membrane penetration. Contributors include John Lenard, Allen Oliff, Richard Schlegel, Patrick R. McClintock, Roland Rueckert, and Michael Rossmann.

Davies, Jeffrey W., ed. *Molecular Plant Virology*. Vol. 1. West Palm Beach, Fla.: CRC Press, 1985.
Davies has assembled an impressive collection of survey papers by leading plant virologists in this resource book on the subject. Scientists such as G. P. Lomonossoff, T. M. A. Wilson, and Michael G. Rossmann describe plant virus vectors, vector interrelationships with host plant cells, the molecular biochemistry of viral infection within host plant cells, lytic and lysogenic cycles, and techniques for inhibiting plant viruses. The various authors provide very clear flowcharts to describe experiments and excellent diagrams of virus DNA and RNA.

Davies, Jeffrey W., et al., eds. *Virus Replication and Genome Interactions*. *Journal of Cell Science*, supp. 7. Cambridge, England: The Company of Biologists, 1987.
A number of prominent molecular biologists and virologists have contributed to

this up-to-date, informative discussion of viruses and their effects on genes within cells. The papers originally were presented at the seventh John Innes Institute Symposium in September, 1986. The evolution of viruses, oncogenic viruses such as Hepatitis B, and the Human Immunodeficiency Virus are among the included topics. Contributors include Arthur Kornberg, Peter H. Duesberg (who disputes HIV as the cause of AIDS), D. B. Stern, and D. J. McGeoch.

Denhardt, D. T., D. Dressler, and D. S. Ray, eds. *The Single-Stranded DNA Phages*. Cold Spring Harbor, N.Y.: Cold Spring Harbor Laboratory, 1978.
This seven-hundred-page work contains extensive survey and research papers that describe bacterial viruses whose genetic material is single-stranded DNA, unlike the normal double-stranded DNA form found in all living cells. Contributing scientists include P. J. Weisbeek, G. N. Godson, D. Dressler, Joachim Messing, and many others. Single-stranded DNA bacteriophage that are discussed in the numerous articles include OX174 and M13. The book includes an extensive author and reference index.

DeVita, Vincent T., Jr., Samuel Hellman, and Steven A. Rosenberg. *AIDS: Etiology, Diagnosis, Treatment, and Prevention*. Philadelphia: J. B. Lippincott, 1988.
This compendious volume, written by three leading experts, describes with great clarity and thoroughness the current scientific understanding of the Human Immunodeficiency Virus (HIV) and the disease for which it is responsible: Acquired Immune Deficiency Syndrome (AIDS). The structure of the virus, its life cycle within infected human T lymphocytes, theories describing how the virus may evade the host's immune system, and whole-body effects upon the infected individual are discussed. Graphic, informative photographs and data tables strongly amplify the importance of this tremendous work.

Diener, T. D. *Viroids and Viroid Diseases*. New York: John Wiley & Sons, 1979.
Diener's excellent introduction to viruses is easy to follow, although the covered material can be deep in parts. Diener describes the structures and mechanisms of infection of viruses, in the process citing evidences from many different viral species. Plant viruses such as tobacco mosaic virus are presented in chapter 2, "Natural Plant Viroid Diseases." The physical and chemical properties of viruses are described in chapters 7 and 8, respectively. Chapter 11, "Nature of Viroids," includes a subsection that considers the possible origins of viruses.

Dimmock, N. J., and S. B. Primrose. *Introduction to Modern Virology*. 1974. 3d ed. Boston: Blackwell Scientific Publications, 1987.
In this important introduction to viruses for the biology student, Dimmock and Primrose explore the various types of viruses, their mechanisms of spread, their effects upon the cells that they infect, and the diseases caused by specific viruses.

The assorted types of both RNA and DNA viruses are described in detail, including viruses such as the Human Immunodeficiency Virus, the cause of Acquired Immune Deficiency Syndrome in humans. The lytic and lysogenic life cycles of viruses within cells is described with great clarity. Chapter 10, "The Evolution of Viruses," speculates upon the origin and development of viruses during the history of life.

Dubow, Michael S., ed. *Bacteriophage Assembly*. New York: Alan R. Liss, 1981.
In this important and very descriptive, molecular study of virus structure and formation within infected host bacterial cells, Dubow and other virologists analyze the biochemical steps leading to viral production and proliferation. Genetic controls, transcription of RNA from DNA, and translation of protein from RNA during the viral lytic cycle of many bacterial virus species are among the topics discussed. Contributors to the volume include Costa Georgopoulos, Isao Katsura, Susan Bock, Helen R. Revel, and Peter Model.

Fermer, F., and A. Gibbs, eds. *Portraits of Viruses: A History of Virology*. New York: Karger, 1988.
Fermer and Gibbs have assembled an important collection of summary papers that describe the history of viral research from the first virus, Tobacco Mosaic Virus, discovered by Dimitri Ivanowski in the early 1890's, to the present. The clear papers also address the possible origins and evolution of viruses as cellular parasites. F. M. Burnet describes "Influenza Virus A," Norton Zinder surveys the "RNA Phage," P. Wildy examines the various "Herpesvirus," and J. Svoboda explores the avian oncogenic virus "Rous Sarcoma Virus."

Fraenkel-Conrat, Heinz. *The Chemistry and Biology of Viruses*. San Diego: Academic Press, 1969.
Fraenkel-Conrat, a leading virologist, presents a concise and sophisticated analysis of viruses and other intracellular parasites in the introduction to the subject. He discusses the possible origins and evolution of viruses, including their species-specificity. Their reproductive cycle, termed lytic, within the cell is described in detail, including all chemical aspects known concerning the mechanisms by which the infecting virus effects a takeover of the cell. The dormant lysogenic cycle is described as well, a situation in which the infecting virus nucleic acid inserts itself into the host-cell DNA. Virus chemical interactions with host-cell membranes are examined also.

Fraenkel-Conrat, Heinz, and Paul C. Kimball. *Virology*. Englewood Cliffs, N.J.: Prentice-Hall, 1982.
Fraenkel-Conrat and Kimball's work is a relatively up-to-date summary of an expanding scientific discipline for the biology student. They emphasize the genetics and molecular biology of viruses at the cellular level. Virus origins and

evolution are discussed, along with the various types of viruses, such as DNA viruses whose genetic material is DNA, RNA viruses whose genetic material is RNA, and adenoviruses that affect the upper respiratory tracts of animals. The molecular basis of host-cell infection is described in detail, including chemical contact with cell membranes, lytic cycles, and lysogenic cycles.

Gallo, Robert. *Virus Hunting: AIDS, Cancer, and the Human Retrovirus: A Story of Scientific Discovery*. New York: Basic Books, 1991.
Gallo, a leading molecular virologist and codiscoverer, with Luc Montagnier, of the AIDS-causing Human Immunodeficiency Virus (HIV), presents his own, personal account of the discovery of the virus that attacks and destroys the T lymphocytes that coordinate many essential activities of the human immune system. He describes for a general audience the molecular structure of the virus, its genetic material (RNA), its means of transmission, and the biochemical events that occur within the infected cells and body of the human victim. Gallo's account provides an excellent perspective upon one of the most deadly diseases to confront humanity in recent history. Excellent microphotographs of the virus accompany Gallo's clear text.

Gallo, Robert C., and Flossie Wong-Staal, eds. *Retrovirus Biology and Human Disease*. New York: Marcel Dekker, 1990.
In this important overview of the molecular basis behind Acquired Immune Deficiency Syndrome (AIDS), Gallo and Wong-Staal describe the Human T-Lymphocyte Virus (HTLV) family of viruses and the Human Immunodeficiency Virus (HIV), the latter being the causative agent of AIDS. Gallo and Luc Montagnier are the codiscoverers of the HIV virus. Contributing authors to this molecular study of the HIV virus include Gallo, William Hardy, James Mullins, William Blattner, Dani P. Bolognesi, Mitsuaki Yoshida, Motoharu Seiki, and R. Jan Gurley.

Gibbs, A. J., ed. *Viruses and Invertebrates*. London: North-Holland, 1973 (distributed by American Elsevier, New York).
Gibbs and several prominent animal/plant virologists examine the infection of host animal and plant cells by various species of virus. In essay two, Kenneth M. Smith examines "Insect Viruses." Marion A. Watson studies "Plant Viruses" in essay three. D. J. Hooper presents viruses that attack segmented worms in essay twelve, "Nematodes." Very clear discussions and excellent microphotographs accompany the texts for each scientific paper. The essays illustrate the similar infective nature of viruses throughout the plant and animal kingdoms.

Gibbs, A. J., and Bryan Harrison. *Plant Virology: The Principles*. New York: John Wiley & Sons, 1976.
Gibbs and Harrison's book is a simple, tremendously clear introduction to the

viruses that parasitize the cells of plant species. They make complicated topics very understandable for the beginning student. They describe experimental methods in virology research, DNA and protein gel electrophoresis of viral molecules, the ecology of viruses and their host cells, and the possible origins of viruses. Excellent viral microphotographs support the text.

Glover, D. M., and B. D. Hames, eds. *Oncogenes*. New York: Oxford University Press, 1990.
Oncogenes are cancer-causing genes, genes whose RNA and/or protein products transform normal, differentiated cells into rapidly multiplying, dedifferentiated cells. Such oncogenes are found within the cells of most prokaryotic and eukaryotic species, lying dormant until some stimulus activates them. Certain viruses (such as Hepatitis B) also carry oncogenes within their genetic material and, therefore, can convert infected cells into cancer cells. In this volume, C. J. Marshall, T. H. Rabbitts, and other scientists contribute papers that analyze and describe the genetics and biochemistry of oncogenes.

Goyal, Sagar M., Charles P. Gerba, and Gabriel Bitton, eds. *Phage Ecology*. New York: John Wiley & Sons, 1987.
This important reference work contains summary and innovative research reports from leading bacteriophage virologists, scientists who are exploring the molecular biochemistry and genetics of bacterial viruses. The contributed papers describe the environmental conditions that favor bacteriophage infection of host cells, the life cycles of bacteriophage, and gene control mechanisms that contribute to the success of viral infections. Contributors of these clear, scientific papers include Donna H. Duckworth, J. N. Coatzee, and Mary Ellen Sanders.

Gust, Ian D., and Stephen M. Feinstone. *Hepatitis A*. West Palm Beach, Fla.: CRC Press, 1988.
Gust and Feinstone's work is an exemplary study of the Hepatitis A virus, a virus that infects the liver cells of humans and other mammals. Their book is packed with useful information concerning transmission of the virus from host-to-host, epidemiology of the virus, viral molecular structure, lytic and lysogenic cycles within the infected host cell, mechanisms of replication, and so forth. They describe techniques for culturing the virus in laboratory cellular suspensions and the history and effectiveness of biological immunizations against the virus. Good microphotographs support the text.

Hendrix, Roger W., Jeffrey W. Roberts, Franklin W. Stahl, and Robert A. Weisberg, eds. *Lambda II*. Cold Spring Harbor, N.Y.: Cold Spring Harbor Laboratory, 1983.
This seven-hundred-page collection of survey and research papers examines one of the most studied viruses in the history of molecular biology: the bacteriophage

lambda. This heavily biochemical and genetic analysis of this bacterial virus examines its ecology, DNA, lytic and lysogenic cycles of infection, assembly during lytic infections, and gene regulation. Nobel laureate Alfred D. Hershey and W. Dove provide an excellent overview of the phage lambda in "Introduction to Lambda." Other contributors include D. Friedman, M. Gottesmann, and Isao Katsura.

Hughes, Sally Smith. *The Viruses: A History of the Concept*. New York: Science History Publications, 1977.
Hughes's work is an impressive study of the history of viral microbiology from the early 1800's to the present. She begins with an examination of the Cell Theory and the Germ Theory of Disease, the latter being developed in the mid-nineteenth century by Louis Pasteur and others. Hughes discusses Russian microbiologist Dimitri Ivanowski's discovery of the first virus in the early 1890's and subsequent discoveries by other virologists during the early decades of the twentieth century. She also tackles medical approaches to viral diseases, including the Salk and Sabin polio vaccines of the 1950's. Individual scientific biographies and excellent photographs support an interesting text.

Klein, George, ed. *Advances in Viral Oncology*. Vol. 3 in *DNA-Virus Oncogenes and Their Actions*. New York: Raven Press, 1983.
Klein presents an array of important scientific papers dealing with oncogenic (cancer-causing) genes and the specific viruses that carry and/or activate these genes within infected host cells. Jesper Zeuther, Ulf Petersson, and Goran Akusjarvi describe the Epstein-Barr Virus, which causes infectious mononucleosis and a rare form of lymph node cancer called Burkitt's Lymphoma. W. Thomas London discusses the Hepatitis B Virus, an oncogenic virus that can cause liver cancer. Peter W. J. Rigby and David P. Lane examine Simian Virus 40.

Knight, C. A. *Chemistry of Viruses*. New York: Springer-Verlag, 1975.
Knight's detailed reference work is a strong survey of the lives of virus particles, entities that parasitize cells and that have been questioned as truly being living forms. Each virus is an assemblage of protein and nucleic acid. Knight discusses the various types of viruses, their biochemical makeup, their morphology, and the chemical events by which they infect cells and reproduce. Knight also examines the experimental means by which scientists purify viruses and the effects of certain chemical and physical treatments upon various viral types.

Koch, Friedrich, and Gebhard Koch. *The Molecular Biology of Poliovirus*. New York: Springer-Verlag, 1985.
This book is an outstanding discussion of the poliovirus, a DNA virus that attacks the central nervous system of higher mammals such as humans and apes, often paralyzing its victims in the process. The Kochs describe the history of the disease

and its study. They also present a detailed molecular biological analysis of the virus, its structure, its means of replication and spread, and so forth. Both the Salk and the Sabin vaccines against the virus are examined as well as their mechanisms of action in conferring human immunity against polio.

Kohn, A., and P. Fuchs, eds. *Mechanisms of Viral Pathogenesis: From Gene to Pathogen*. Boston: Martinus Ninjhoff, 1983.
Papers from the twenty-eighth OHOLO Conference in Zichron Ya'acov, Israel, are presented in this volume, which summarizes conference proceedings. More than one hundred molecular virologists participated in this meeting of virus experts. Many viral species are described in several dozen papers. Y. Aloni provides a summary of the many different animal viruses. R. M. Elliott and David H. L. Bishop describe the characteristics of bunyaviruses. M. Carter details the biology of measles viruses. The papers contain detailed molecular data.

Kolber, Alan, ed. *Tumor Virus-Host Cell Interactions*. New York: Plenum Press, 1973.
The molecular virology research papers presented in this volume describe the mechanisms by which viruses attach to host-cell membranes, how they insert their genetic information into the infected cells, and how the infecting viral DNA or RNA transforms the cell, thereby commandeering the control mechanisms of the cell for viral use. Various virus species are cited, including adenoviruses, avian oncornaviruses, and herpes viruses. Contributors include Marc Girard, Gerhard Sauer, R. Dulbecco, Bernhard Hirt, and Carol Prines.

Kurstak, Edouard, ed. *Arctic and Tropical Arboviruses*. New York: Academic Press, 1979.
In this volume, Kurstak and other virologists examine the arboviruses, viruses that infect and parasitize many species. The papers included focus upon the molecular biology and patterns of infection of arboviruses that attack species in the colder arctic regions of northern Canada and the Soviet Union and in the warm tropical rain forests of Central and South America, central Africa, and southeast Asia. Other contributors include John Woodall, Wayne Thompson, Victor Stollar, D. K. Lvov, and S. P. Chunikhin.

Kurstak, Edouard, and Karl Maramorosch, eds. *Viruses and Environment*. New York: Academic Press, 1978.
Kurstak and Maramorosch have put together a very interesting collection of survey papers and research articles that address various aspects of viruses and the hosts that they infect, including factors such as cancer incidence, epidemiology, and structural variations within viruses. In part 1, "Viruses, Cancer and Environment," George Klein, Guy de The, Robert Gallo, and others discuss viruses and phage that cause cancer, viruses such as Hepatitis B and Rous

Sarcoma Virus. In part 4, "Viral Variations: Molecular and Epidemiological Aspects," D. K. Lvov, Jay Levy, Edwin Kilbourne, and others examine biochemical changes within various viral types and their effects upon viral infectiveness and disease transmission/incidence.

____________, eds. *Viruses, Evolution, and Cancer: Basic Considerations*. New York: Academic Press, 1974.
This eight-hundred-page volume represents the proceedings of the second International Conference of Comparative Virology, which was held in 1973 in Mont Gabriel, Canada, near Montreal. Clear presentations highlight the informative papers in this thorough panorama of molecular virology. Major topics include embryo viruses, virus interactions with host cells, oncogenic (cancer-causing) DNA viruses, adenoviruses, herpes virus and abnormalities in cellular growth, RNA tumor viruses, and avian (bird) tumor viruses. Contributing scientists include Robert S. Kauffman, George Klein, Alan Liss, Frederick Murphy, Fred A. Rubenstein, Davor Solter, Philippe Vigier, and Elizabeth Szybalski, among many others.

Linke, Paul A., ed. *Viruses and Plasmids in Fungi*. Vol. 1. New York: Marcel Dekker, 1979.
Linke has collected an impressive array of scientific papers that examine the viruses that infect fungal species ranging from yeast to mushrooms. Fungal virus physical structures, principal molecular components, lytic and lysogenic cycles, and effects upon host cells are among the key topics emphasized in the volume. Contributing authors include Kenneth Buck, Robert Bozarth, Diane Mitchell, E. Allan Bevon, Michel Guerineau, and Peter R. Day.

Luria, Salvador E., James E. Darnell, Jr., David Baltimore, and Allan Campbell. *General Virology*. New York: John Wiley & Sons, 1953, 1967, 1978.
This six-hundred-page introduction to viruses for the biology student is an excellent survey of the subject. The authors, who include several Nobel laureates, are leading authorities on virus genetics and biochemistry. They describe, chapter-by-chapter, the various types of viruses, the organisms that these viruses infect, the life cycles of virus particles, and the structures and functions of viruses within infected cells.

McClintock, Barbara. *The Discovery and Characterization of Transposable Elements: The Collected Papers of Barbara McClintock*. Edited by John A. Moore. Genes, Cells, and Organisms: Great Books in Experimental Biology 17. New York: Garland, 1987.
This one-thousand-page work contains reprints of most of McClintock's seventy-three research papers, which date over many decades of the twentieth century. McClintock, winner of the 1983 Nobel Prize in Physiology or Medicine, is the

discoverer of transposable elements, small segments of DNA (genes) that can excise themselves from one chromosomal location and insert themselves into a new chromosomal location elsewhere. Her discovery has made a major impact upon the modern science of molecular biology, particularly with respect to cloning. This book represents a monument to her insight, brilliance, and perseverance as a scientist.

Madeley, C. R., and Anne M. Field. *Virus Morphology*. Edinburgh, Scotland: Churchill Livingstone, 1988.
Distinctive electron microphotographs of viruses and discussions of dozens of viral types highlight this important reference book. Madeley and Field provide a step-by-step presentation of many virus species, particularly those viruses that infect humans. The information presented in the book includes viral life cycles, diseases, and treatments. A perfect reference for the student of virology and/or medicine.

Maramorosch, Karl, and John J. Kelvey, Jr., eds. *Subviral Pathogens of Plants and Animals: Viroids and Prions*. San Diego: Academic Press, 1985.
Richard I. Carp, S. Patricia Cochran, Karl H. J. Gordon, Paul Keese, Michael Kiefer, Roger C. Moretz, and Robert A. Owens are among the many contributors to this collection of survey articles describing noncellular life forms that parasitize cellular life, specifically viruses and prions. Both types of organisms consist of a few molecular constituents, although they very effectively reproduce within host cells, either using the cell (the lysogenic cycle) or destroying the cell (the lytic cycle) in the process. Prions apparently contain no nucleic acid, which would be extraordinary if confirmed. Prions, animal viruses, and plant viruses are described in detail.

Marks, Paul A., ed. *Genetics, Cell Differentiation, and Cancer*. Orlando, Fla.: Academic Press, 1985.
From the seventh Bristol-Meyers Symposium on Cancer Research at the Memorial Sloan-Kettering Cancer Center, this book is a compendium of papers presented at the conference. Many of the world's leading virologists present informative papers in this volume, including J. Michael Bishop, Judith Levin, Tiernko Hanafusa, James E. Darnell, Jr., Susumu Tonegawa, and Robert Weinberg. Major paper topics include the molecular basis of cancer and cellular transformation, the "Z" form of DNA, adenovirus DNA, and tumor viruses.

Matthews, Christopher K., Elizabeth M. Kutter, Gisela Mosig, and Peter B. Berget, eds. *Bacteriophage T4*. Washington, D.C.: American Society of Microbiology, 1983.
In this detailed work, D. Peter Snustad, Nancy G. Nossal, Edmond Brody, Jonathan King, and many other bacterial virologists analyze all aspects of the

bacterial virus bacteriophage T4, a virus that infects the human intestinal bacterium *Escherichia coli* and that is the most studied bacterial virus in the world. The phage T4 viral life cycle, mechanisms of lysis and lysogeny, DNA replication, RNA transcription, protein translation, gene regulation, three-dimensional molecular structure, and genetic cloning are thoroughly examined by the contributing authors to this volume.

Maza, Luis M. de la, and Ellena M. Peterson. *Medical Virology V.* Hillsdale, N.J.: Lawrence Erlbaum, 1986.
This five-hundred-page volume represents the proceedings of the 1985 International Symposium on Medical Virology held in Anaheim, California. The scientific papers and monographs presented in this book cover such topics as Acquired Immune Deficiency Syndrome (AIDS) and the HTLV-III viruses, herpesvirus, and automation experiments to study viruses without endangering medical researchers. Contributing scientists to this important reference work include Robert Gallo, J. W. D. Kay, and Monto Ho.

Monath, Thomas P., ed. *The Arboviruses: Epidemiology and Ecology*. Vols. 1-5. Boca Raton, Fla.: CRC Press, 1988.
This five-volume compendium of research and survey articles summarizes practically everything that is known about arboviruses in slightly more than one thousand pages. In chapter 1, Paul Bres describes the "Impact of Arboviruses on Human and Animal Health." In chapter 3, "Virus Variation and Evolution: Mechanisms and Epidemiological Significance," Barry J. Beaty, Dennis W. Trent, and John T. Roehrig examine the molecular changes in virus structure and the evolution of different viral types to enhance the spread of certain arbovirus diseases. Thomas W. Scott addresses arbovirus interactions with animal hosts in chapter 10, "Vertebrate Host Ecology." Arboviruses are described group-by-group in several papers. Other contributors include Patricia H. Webb, Thomas Monath, and Robert B. Tesh.

Nelson, Oliver, ed. *Plant Transposable Elements*. New York: Plenum Press, 1988.
Nelson's book, which is dedicated to the late geneticist Alexander Hollaender, who died in December, 1986, is a collection of research papers describing movable genetic elements called transposons. Transposons, also called transposable elements, are segments of DNA containing one or a few genes that can relocate from one portion of a chromosome to another. Some scientists have hypothesized that transposons may be the evolutionary ancestor of viruses. Contributors to this important volume include N. Federoff, Peter A. Peterson, Rosemary Carpenter, Virginia Wolbot, Brian Charlesworth, Oliver Nelson, Karen Dishi, and Vicki Chandler.

Nicolau, Claude, ed. *Virus-Transformed Cell Membranes*. New York: Academic Press, 1978.

In this detailed, but clear, work, Nicolau presents six summary research papers by leaders in the study of virus-host cell membrane research. The chemistry of cell surface membranes from various species are described, particularly with reference to receptor proteins that interact with viral proteins during the initial infection process. Subsequent cascade chemical events in the infected cell membrane and within the cell are described as well. Contributors to this volume include Nicolau, Kenneth Noonan, Fred Rapp, Sheena Johnston, Marion R. Steiner, and Sheldon Steiner.

Notkins, Abner Louis, and Michael B. A. Oldstone, ed. *Concepts in Viral Pathogenesis II*. New York: Springer-Verlag, 1986.

Notkins and Oldstone have assembled a unique collection of clear but very chemically oriented research papers that deal with viruses and their distinct mechanisms of cellular infection. Contributors include Edwin D. Kilbourne, Fred Brown, David H. L. Bishop, Ian A. Wilson, and Heiner Westphal. They discuss various viral topics such as the structures of viruses, oncogenes (cancer-causing genes), which are associated with certain viruses such as Hepatitis B, cellular transformation during viral infection, human and animal immunity against viruses, vaccines and antiviral chemicals aimed at destroying viruses.

Paste, George, and Garth L. Nicolson, eds. *Virus Infection and the Cell Surface*. New York: North-Holland, 1977.

Paste and Nicolson have put together an interesting collection of scientific papers that attempt to determine the precise molecular mechanisms by which viruses attach to, and initiate infection of, host cells. Chemical and physical interactions between virus coat protein and proteins/lipids within the host-cell membrane receive primary experimental focus. Several virus case studies are considered, including Sendai virus and bacteriophage. Contributors to the volume include Thomas Bachi, William Burns, Jane E. Deas, and Richard T. Johnson.

Pollard, Morris, ed. *Perspectives in Virology*. Vol. 10. New York: Raven Press, 1978.

Pollard has assembled an impressive collection of presentation papers made at the 1977 Gustav Stern Symposium. Each paper represents a summary of major research for specific viral types. Many virus species are discussed, although the more recently diagnosed viruses of the 1980's (such as HIV and Hepatitis C) are not included. Among the symposium presenters include Norton D. Zinder, who describes the biology of the F1 bacteriophage, R. L. Crowell, who describes the coxsackie family of viruses, and Peter C. Doherty, who studies human immunity to viral infections. Jordi Cassals explores "New Virus Diseases of Man," such as lassa fever.

Proffitt, Max R., ed. *Virus-Lymphocyte Interactions: Implications for Disease*. New York: Elsevier/North-Holland, 1979.
The collection of scientific reports found in this book are almost prophetic, since only a few years were to pass before the AIDS pandemic, which is caused by a virus that attacks human immune lymphocytes, became pronounced on a global scale. The scientific reports in this volume were presented at the second Immunodynamics Conference at the Cleveland Clinic in June, 1979. While specifically not geared toward AIDS, the presenters look at the cellular mechanisms (such as lymphocytes) by which the human body defends itself from viral infections. Contributors include Murray B. Gardner, J. Edwin Blalock, Jeffrey J. Collins, and Anthony J. Dennis. Viral diseases such as herpes, measles, and cytomegalovirus are examined in the various reports.

Prusiner, Stanley B., and Michael P. McKinley, eds. *Prions: Novel Infectious Pathogens Causing Scrapie and Creutzfeldt-Jacob Disease*. San Diego: Academic Press, 1987.
Prusiner and McKinley have assembled an important collection of research papers on an apparently new form of life, the prion, which is viruslike in its mechanism of action but which seems to have no DNA. It may rely upon the DNA of its host cells for its genetic material in some mysterious fashion. Prions are associated with certain mammalian and human degenerative central nervous system diseases, such as Kuru, scrapie, and Creutzfeldt-Jacob Syndrome. Prusiner authors several papers in this volume, as does McKinley. Other contributors include Tikvah Alper, Darlene F. Groth, Karen A. Bowman, and David T. Kingsbury.

Ptashne, Mark. *A Genetic Switch: Gene Control and Phage Lambda*. Cambridge, Mass.: Blackwell Scientific Publications/*Cell* Press, 1985.
Ptashne, a leading bacteriophage genetics researcher, examines the complicated genetics and biochemistry of the bacteriophage lambda, a virus that infects cells of the human intestinal bacterium *Escherichia coli* and that is one of the most studied organisms in the history of molecular biology, in this short work. Ptashne writes with great clarity in describing the molecular structure of the virus, control of genetic expression in the virus, and genetic regulation of the virus' lytic and lysogenic infectious cycles. Beautifully clear diagrams and microphotographs help to clarify major points in this powerfully descriptive study.

Radetsky, Peter. *The Invisible Invaders: The Story of the Emerging Age of Viruses*. Boston: Little, Brown, 1991.
Radetsky's outstanding work is one of the few books on virology that is written specifically for the layperson. Rodetsky provides a very clear and comprehensive coverage of the history of virology, the various classifications of viruses, their means of transmission, their host specificities, and their molecular structures. He discusses certain viruses of interest, such as the many different varieties of cold-

producing rhinoviruses in humans and the cancer-causing (oncogenic) Hepatitis B Virus. He also discusses viral vaccines and gene therapy (cloning) using modified viruses.

Riley, Matilda White, Marcia G. Ory, and Diane Zablotsky, eds. *AIDS in an Aging Society: What We Need to Know*. New York: Springer, 1989.
In this important reference book devoted to the deadly AIDS pandemic, Riley, Ory, and Zablotsky have compiled several papers that describe current research into the Human Immunodeficiency Virus. The included papers deal with the life cycle of the virus, methods of preventing the disease's spread, methods of attacking the disease, and the disease's impact upon society.

Robertson, Hugh D., Stephen H. Howell, Milton Zaitlin, and Russell M. Malmberg, eds. *Plant Infectious Agents: Viruses, Viroids, Virusoids, and Satellites*. Cold Spring Harbor, N.Y.: Cold Spring Harbor Laboratory, 1983.
The prestigious Cold Spring Harbor Biological Laboratory on New York's Long Island was the site of this 1983 plant virology conference in which several hundred scientists participated. Robert Goodman, Timothy C. Hall, Kenneth Buck, Andrea Branch, and Thomas J. Morris head an impressive list of plant virologists who presented papers at the conference. Major topics included the various types of DNA and RNA plant viruses, plant viral vectors, mechanisms of infection, lytic and lysogenic cycles of plant viruses, and plant tumor viruses.

Roth, Joy S. *All About AIDS*. New York: Harwood Academic Press, 1989.
Roth's concise work is a thorough and straightforward description of Acquired Immune Deficiency Syndrome for the layperson. The Human Immunodeficiency Virus is described, along with its structure, means of transmission, life cycle within host cells, and its overall impact upon the infected person. Research into the disease is discussed along with the social impact presented by this virus.

Schupbach, Jorg. *Human Retrovirology: Facts and Concepts*. New York: Springer-Verlag, 1990.
This short, clear book, written by a physician, is a fascinating history of retroviruses, viruses whose genetic material is RNA and that use an enzyme called reverse transcriptase to reproduce viral DNA inside host cells. Schupbach concentrates upon specific retroviral types, in particular the Human Immunodeficiency Virus (HIV), the cause of Acquired Immune Deficiency Syndrome (AIDS). He describes the molecular structure, means of transmission, and biochemistry of replication of these viruses within their host cells. Twenty-four figures accompany the text.

Scott, Andrew. *Pirates of the Cell: The Story of Viruses from Molecule to Microbe*. New York: Basil Blackwell, 1985.

Scott's extremely clear presentation of viruses is one of the few books on the subject that is written explicitly for the layperson. He devotes several chapters to such viral topics as viral structure (DNA wrapped within a protective protein sheath), viral invasiveness with respect to host-cell membrane penetration, viral reproduction (lytic cycle) within the host cell, human diseases caused by viruses, cancer-causing (oncogenic) viruses, antiviral vaccines, and human exploitation of viruses for projects such as molecular cloning of genes into other organisms. Scott covers all angles of the subject very comprehensively.

Shapiro, James A., ed. *Mobile Genetic Elements*. New York: Academic Press, 1983.
Shapiro has compiled a collection of major scientific papers by leading researchers in molecular genetics, particularly experts on transposable elements, segments of genes that "jump" from one chromosomal location to another, in the process often affecting gene regulation. Contributors include Nina V. Federoff, Allan Campbell, Nancy Kleckner, Gerald M. Rubin, Margaret G. Kidwell, Harold E. Varmus, and P. Borst. They describe the activities of "transposons" in a variety of species such as maize, bacteriophage, fruit flies, retroviruses, and trypanosomes.

Smith, Kenneth M. *Plant Viruses*. London: Chapman & Hall, 1935. 6th ed. New York: John Wiley & Sons, 1977.
Smith's clear introduction to the study of plant viruses is a thorough presentation of a complex subject for the beginning student. He describes the physiology and biochemistry of plant viruses, including their means of transmission, their effects upon hosts, and their routes of infection within host cells. Smith also divides the various types of viruses into various categories and describes experimental techniques for culturing plant viruses in cellular suspensions in the laboratory.

Smith, Kenneth M. *Virus-Insect Relationships*. New York: Longman, 1976.
In this excellent survey work, Smith discusses the various viruses that are known to infect and parasitize the cells of insect species. He also addresses the evolutionary relationship by which some insect species (such as mosquitoes) can transport viruses to specific mammalian hosts without themselves being affected. As such, insects are common vectors of many viral diseases. Smith's book considers both inclusive and noninclusive viral diseases. Excellent photographs and microphotographs supplement the text.

Stent, Gunther, selector. *Papers on Bacterial Viruses*. Boston: Little, Brown, 1960, 1965.
In this sequel to Joshua Lederberg's *Papers in Microbial Genetics: Bacterial and Bacterial Viruses*, Stent presents major scientific breakthrough papers from the explosive era of molecular biology in the 1950's and early 1960's. Research papers describing the nature, origins, and structures of viruses are presented,

many by Nobel laureates. The included reprint articles are by such eminent scientists as Alfred Hershey, Martha Chase, James D. Watson, Olae Maaloe, Seymour S. Cohen, Arthur Kornberg, Salvador Luria, George Streisinger, Francis Crick, Seymour Benzer, Gunther Stent, Andre Lwoff, Francois Jacob, Leslie Barnett, and Matthew Meselson. Thirty papers are presented in this four-hundred-page compilation.

Stephenson, John R., ed. *Molecular Biology of RNA Tumor Viruses*. New York: Academic Press, 1980.
Stephenson's work contains scientific essays written by leading virologists and researchers in the field of RNA tumor viruses, viruses whose genetic material is RNA and that cause tumor growth in infected cells of the animals and plants that they attack. In chapter 1, Murray B. Gardner provides a "Historical Background" into RNA tumor virus research, including viruses such as avian sarcoma virus and mouse mammary tumor virus. In chapter 2, George J. Todaro describes "Interspecies Transmission of Mammalian Retroviruses." Gary F. Gerard and Duane P. Grandgenett describe the RNA-to-DNA copying viral enzyme reverse transcriptase in chapter 9, "Retrovirus Reverse Transcriptase."

Stuart-Harris, Sir Charles H., and C. W. Potter, eds. *The Molecular Virology and Epidemiology of Influenza*. San Diego: Academic Press, 1984.
This excellent survey work is a transcript of papers presented at the sixth Beecham Colloquium on Aspects of Infection. Sir Christopher Andrewes, G. C. Schild, H. Smith, Edwin D. Kilbourne, and several other leading virologists describe the biology and spread of the influenza viruses that still are major killers in many portions of the world. Major topics include the influenza A type viruses, global pandemics of influenza during recent history, and the human immune responses against influenza infections. Good diagrams enhance the contributed papers.

Tooze, John, ed. *The Molecular Biology of Tumor Viruses*. Cold Spring Harbor, N.Y.: Cold Spring Harbor Laboratory, 1973.
Tooze's book is a large assembly of research papers that analyze the genetics and biochemistry of tumor viruses, both DNA and RNA viruses, which are both benign and malignant in their effects. RNA tumor viruses such as Rous sarcoma virus (the first discovered cancer virus) and avian leukemia virus are examined by Tooze and colleagues. DNA tumor viruses such as polyoma virus and the Epstein-Barr Virus (cause of infectious mononucleosis and Burkitt's lymphoma) are also considered in numerous papers. Chapter 2, "The Culture of Mammalian Cells," describes how viruses are studied with host mammalian cells in the laboratory.

Trotot, Pierre M. *Imaging of AIDS*. Philadelphia: B. C. Decker, 1991.
Trotot's work is a tremendous sourcebook of information concerning the Human Immunodeficiency Virus (HIV) and the disease that it causes, Acquired Immune Deficiency Syndrome (AIDS). The virus, which presumably evolved from less-pathogenic lymphocyte viruses during the 1960's and 1970's, is described in considerable detail. Many powerful electron microphotographs of the virus and photographs of infected body regions make this book an important reference for the student of virology, for physicians, and for health professionals.

Waterson, A. P., and Lise Wilkinson. *An Introduction to the History of Virology*. New York: Cambridge University Press, 1978.
Waterson and Wilkinson's book is an excellent discussion of viruses and the history of their study for the layperson. They describe the history of virus research during the nineteenth and twentieth centuries, the various types of DNA and RNA viruses that infect various species, and the overall mechanisms of transmission and infectivity of host cells by viruses. They describe bacterial viruses (bacteriophage) and tumor viruses (oncophage). Experimental methodologies and cell culturing of viruses in the laboratory are discussed as well.

Wolstenholme, G. E. W., and Maeve O'Connor, eds. *Strategy of the Viral Genome*. Edinburgh, Scotland: Churchill Livingstone, 1971.
Mark Ptashne, David Baltimore, William Erkhart, and J. H. Subak-Sharpe are among the many virologists who present research papers in this volume, a collection of papers from a CIBA Foundation Symposium on the molecular biology of viruses. The genetics and biochemistry of many different virus types are presented in these papers. Viruses studied include adenovirus, Simain Virus 40, poliovirus, and bacteriophage.

Zinder, Norton, ed. *RNA Phages*. Cold Spring Harbor, N.Y.: Cold Spring Harbor Laboratory, 1975.
In this important survey work, Zinder and numerous virologists analyze the molecular biochemistry of RNA phages, viruses whose genetic material is RNA, not DNA. This work emphasizes bacterial viruses, specifically the bacteriophage. RNA phage mechanisms of infection, replication within their host cells, and physical structures are presented. Contributing scientists to this volume include Walter Konigsberg, H. D. Robertson, Joan A. Steitz, H. F. Lodish, L. Shapiro, and N. Fedoroff.

Chapter 8
THE EUKARYOTES

Symbiotic Origins

Bellig, Robert, and George Stevens, eds. *Nobel Conference XXIII: The Evolution of Sex*. San Francisco: Harper & Row, 1988.
This special Nobel conference, held at Gustavus Adolphus College, brings together prominent scientists, philosophers, and theologians to discuss the importance of sexual reproduction in the evolution of eukaryotic life on earth. J. M. Smith summarizes "The Evolution of Sex." Lynn Margulis and Dorion Sagan present "Sex: The Cannibalistic Legacy of Primordial Androgynes." Peter H. Raven describes special evolutionary adaptations (such as flowers) for sex in plants. William Donald Hamilton speculates upon resistance to disease as a result of sex. Sarah Blaffer Hardy discusses "The Primate Origin of Human Sexuality." Philip J. Hefner presents "Sex, for God's Sake: Theological Perspectives." Panel discussions follow each article.

Brues, Austin M., and George A. Sacher, eds. *Aging and Levels of Biological Organization*. Chicago: University of Chicago Press, 1965.
This volume, dedicated to Henry Quastler, is a transcript of scientific research presentations and panel discussions at an aging research conference held at Princeton University in December, 1962, under the auspices of the American Institute of Biological Sciences, the National Heart Institute, and the National Institutes of Health. Among the conference participants were Austin Brues, Theodosius Dobzhansky, James Birren, and Charles Ehret. They discuss many diverse topics, including genetic and environmental influences upon the aging of eukaryotic organisms, the roles of hormonal and circadian rhythms in aging processes, and even the extreme longevity of early Old Testament patriarchs.

Cristofalo, Vincent J., and Emma Holeckova, eds. *Cell Impairment in Aging and Development*. New York: Plenum Press, 1975.
Based upon scientific papers presented at a 1974 symposium at the Castle of Hruba in Skola, Czechoslovakia, this book is a lengthy discussion of senescence (aging), DNA and gene regulation during organismal development, chemicals and molecules that contribute to or hinder aging-related processes, and the aging of laboratory-grown cell cultures. Good microphotographs and data tables supplement the presentation papers. Contributors to the volume include Cristofalo, L. M. Franks, Robert J. Hay, G. M. Martin, Robert T. Dell'Drco, and William D. Barnes.

Danon, David, N. W. Shock, and M. Marois, eds. *Aging: A Challenge to Science and Society*. New York: Oxford University Press, 1981.
This collection of scientific articles and presentations was assembled from a 1980 aging research conference held at L'Institute de la Vie in Vichy, France, a conference attended by more than two hundred scientists. Leaders in the field of aging research discuss major advances in the field. Leonard Hayflick and Curtis Sprague examine aging in cell populations. Harold Brody and David Danon look at nervous system aging. George Sacher and N. W. Shock speculate upon the possibilities of life extension for humans. Other contributors include Robert Terry, Nicolae Simonescu, James L. Hodgson, and Richard A. Rifkind.

Finch, Caleb E. *Longevity, Senescene, and the Genome*. Chicago: University of Chicago Press, 1990.
Finch, a University of Southern California gerontologist, provides a fascinating new glimpse at why and how living organisms, especially eukaryotes, age. He challenges accepted theories that limit species life spans based upon very few case studies. He presents a new theory, supported by multispecies mortality rate constants, in which the process of aging in organisms is based upon genetics and environmental influences that have been shaped during millions of years of evolutionary change within the organism's species.

Ghiselin, Michael T. *The Economy of Nature and the Evolution of Sex*. Berkeley: University of California Press, 1974.
Ghiselin's philosophical, psychological, and evolutionary look at the origins and nature of sexual reproduction in eukaryotic species is a challenging approach to the subject. He describes reproductive cells (gametes) and the process of meiosis in chapter 3, "The Loves of the Plants, or the Biological Role of Sex." He relates male animal aggression and sex in chapter 5, "The Copulatory Imperative, or, Male Combat." He continues with this discussion in chapter 6, "Seduction and Rape, or, Female Choice and Male Sequestering."

Hall, Brian K. *The Neural Crest*. New York: Oxford University Press, 1989.
Hall's book is actually two books in one. It includes a reprint of neurophysiologist Sven Horstadius' *The Neural Crest: Its Properties and Derivatives in the Light of Experimental Research*. Hall then describes current progress in vertebrate neurological research as relating to the neural crest cells, a set of differentiated ectodermal cells that are highly plastic and that are responsible for the extensive development of the vertebrate animal nervous system, the most complex nervous system in any group of animals. Mammalian intelligence eventually can be traced back to the origins of neural crest tissue.

Inoue, Shinya, and Raymond E. Stephens, eds. *Molecules and Cell Movement*. Vol. 30 of the Society of General Physiologists series. New York: Raven, 1975.

Based upon a 1974 cell biology symposium, this book is a collection of presentation papers dealing with subcellular and cellular motion, including the movements of molecules, particularly within eukaryotic cells. Covered topics include chromosomes, the mitotic separation of chromosomes prior to cell division, ciliary beating, the rotation of flagella within the cell membrane, amoeboid movement in amoebae and other species, and the important role of the molecule actin in animal muscle contraction. Contributors to this volume include Inoue, E. W. Taylor, R. Bruce Niklas, Kenneth A. Johnson, Peter Satir, Roger Eckert, Hope Ritter, and R. Rappaport.

Kroon, A. M., and C. Saccone, eds. *The Biogenesis of Mitochondria: Transcriptional, Translational, and Genetic Aspects*. New York: Academic Press, 1974.
Kroon and Saccone have assembled scientific papers from numerous prominent cell and evolutionary biologists into this five-hundred-page compendium. The contributed papers describe the genetics, origins, and evolution of mitochondria, components of all living eukaryotic cells that are the sites of energy production within these cells. Mitochondria are interesting in that they contain their own genetic material and reproduce independently of the cell, phenomena that indicates an independent origin. Contributors to this volume include I. B. Dawid, G. J. Wu, D. E. Griffiths, P. Borst, C. S. Cooper, and C. J. Avers.

Lloyd, Clive W., and David A. Rees, eds. *Cellular Controls in Differentiation*. New York: Academic Press, 1981.
This book is a transcript of a December, 1980, biochemistry symposium held in The Netherlands to recognize the fiftieth anniversary of the Unilever Company. The keynote address, "Genes and Development," by Sydney Brenner introduces an important topic that serves as the basis of life, evolution, and individual development. Klaus Weber and Mary Osborn describe "Aspects of the Cytoskeleton of Mammalian Cells," Philip Coffino presents a "Genetic and Biochemical Analysis of Regulation by Cyclic AMP," and Arthur B. Pardee and Ruth Sager illustrate "Cancer as Genetically Defective Differentiation." All scientific papers in this volume conclude with a question-and-answer session. Other important contributions come from Donald Metcalf, Natalie M. Teich, Janice Rowe, and Piet Borst.

Margulis, Lynn. *Origin of Eukaryotic Cells: Evidence and Research Implications for a Theory of the Origin and Evolution of Microbial, Plant, and Animal Cells on the Precambrian Earth*. New Haven, Conn.: Yale University Press, 1970.
This fantastic book by a brilliant evolutionary scientist summarizes her own daring proposal that mitochondria, critical components of all eukaryotic cells, have an independent origin as single cells more than one-and-a-half billion years ago. In chapter 1, she presents "Problems of Cellular Evolution," such as the presence of DNA in mitochondria and chloroplasts. In chapter 2, she propounds her

"Symbiotic Theory" by which mitochondria and chloroplast ancestor cells were incorporated into, and utilized by, host cells, thereby giving rise to the first eukaryotes. She provides much evidence to support this once-criticized and now-accepted theory.

Rose, Michael R. *Evolutionary Biology of Aging*. New York: Oxford University Press, 1990.
Rose's clear presentation of aging, a major phenomenon in eukaryotes, includes a detailed account of evolutionary research into the process of senescence (aging). Rose presents a position by which aging occurs as a relaxation of organismal self-maintenance coincident with the beginning of the organism's sexual reproductive phase. The organism subsequently deteriorates following this stage. Rose presents the various theories of aging, including the molecular clock and free radical hypotheses, in comparison to his own proposal. He argues that human life span can be drastically extended by current molecular evolutionary research.

Smith, John Maynard. *The Evolution of Sex*. New York: Cambridge University Press, 1978.
In this insightful work, the eminent evolutionary scientist Maynard Smith examines the evolution of sexual reproduction, one of the predominant characteristics of most eukaryotic species, from both genetic and behavioral viewpoints. Chapter 2, "Some Consequences of Sex and Recombination—I. The Rate of Evolution," links sexual reproduction, the occurrence of mutation, and natural selection for or against various mutations. In chapter 6, Smith examines "Short-Term Advantages for Sex and Recombination—I. An Unpredictable Environment." He considers "female choice," parental care, monogamy, and polygamy in chapter 10, "Sexual Selection."

Strehler, Bernard L. *Time, Cells, and Aging*. New York: Academic Press, 1977.
In this comprehensive book, Strehler presents a detailed survey of recent research and theories regarding senescence, the phenomenon of aging in animals. Chapter 3, "The Distribution of Cellular Aging," includes sections on "Nonaging Systems," such as reproductive cells, and "On Species Which May Not Age," such as cellular slime molds. Chapter 6, "Proximate Effects of Cellular Aging," describes the physiological decline of individual organisms as they grow older, while chapter 9, "Theories on the Mechanisms of Aging," looks at the probable causes of aging in living systems. The book is very informative and includes many mathematical models of aging causes and effects.

Truman, D. E. S. *The Biochemistry of Cytodifferentiation*. Oxford, England: Blackwell Scientific Publications, 1974.
Truman, a geneticist at the University of Edinburgh, discusses the elaborately evolved processes of cellular development, differentiation, and development within

the individual organism. The short book begins with a discussion of theory in chapter 1, "The Biology of Cell Differentiation." Chapters 3, "The Regulation of Cellular Activity," and 4, "The Genes and the Tissues," examine the molecular control mechanisms that contribute to specialization of organismal structures. Chapter 6, "Some Systems of Cytodifferentiation," examines differentiation of skin, liver, muscle, eye lens cells, and red blood cells. The Britten and Davidson model of gene regulation in development is also discussed.

Woodhead, Avril, and Keith H. Thompson, eds. *Evolution of Longevity in Animals: A Comparative Approach*. New York: Plenum Press, 1987.
This important volume, dedicated to the late George Sacher, is a compilation of essays dealing with scientific research into the causes of senescence (aging), a phenomenon that is predominant in most eukaryotic species. Robert Arking discusses "Genetic and Environmental Determinants of Longevity in *Drosophila*." Richard L. Russell and Renee I. Seppa examine "Genetic and Environmental Manipulation of Aging in *Caenorhabditis elegans*." Henry R. Hirsch asks, "Why Should Senescence Evolve?" Thomas E. Johnson and colleagues look at "Mutant Genes That Extend Lifespan." Alan R. Templeton and colleagues describe "The Proximate and Ultimate Control of Aging in *Drosophila* and Humans."

Animalia, Plants, and Fungi

Adamson, Joy. *Born Free*. New York: Pantheon Books, 1970.
Adamson's best-seller, the story of her unique interaction with the lioness Elsa, inspired several films, subsequent books, and worldwide concern for animal rights. Her book describes the raising of Elsa and her siblings from young cubs up to adulthood. It is an excellent chronicle of lion behavior. Adamson and her husband's successful efforts to return Elsa to the wild represent an inspiring portion of the book, which includes numerous candid photographs.

____________. *The Spotted Sphinx*. New York: Harcourt, Brace & World, 1969.
Adamson, famous for her study of the lioness Elsa in *Born Free*, describes the life of the female cheetah Pippa during the three-year period from 1964 to 1967. She traces the experiences of Pippa and her young in the wild, providing a unique glimpse of this threatened species of the African savanna. Excellent color photographs supplement the text.

Ainsworth, G. C., and Alfred S. Sussman, eds. *The Fungi: An Advanced Treatise*. Vols. 1-2. New York: Academic Press, 1965.
Ainsworth and Sussman have assembled an outstanding collection of survey papers that describe various aspects of one of the major groups of eukaryotic organisms on earth, the fungi. Fungi serve as decomposers in the recycling of matter through

living ecosystems. The two volumes of this work total approximately fifteen hundred pages and include articles by Ainsworth, Richard Morita, Allan Campbell, D. G. Catcheside, and C. F. Robinow. Excellent diagrams of fungi accompany each article. Several articles are devoted to fungal genetics, nutrition, and antibody production, the latter being of interest to medical research.

Alexander, R. McNeill. *Optima for Animals*. London: Edward Arnold, 1982.
Using mathematical models where appropriate, Alexander, a professor at the University of Leeds, describes the evolution of various specialized physical adaptations within the animal kingdom. He examines such things as the development of compound eyes in insects, bone strength in vertebrate animals, mechanisms of flight in birds and in other species, and locomotion and motion in terrestrial animals. He also examines various aspects of physiology and behavior such as reproductive cycles, hunger and thirst, and territorial behavior. The book is very comprehensive and informative in its coverage and scope.

Anderson, Sydney, and J. Knox Jones, Jr. *Recent Mammals of the World: A Synopsis of Families*. New York: Reinhold Press, 1967.
Sponsored by the American Society of Mammalogists, this work is a comprehensive survey of mammalian species worldwide for both the layperson and the expert. The authors proceed through the various mammalian species family-by-family. Each species is described according to anatomical structures, behavior, basic characteristics, and worldwide distribution. The work includes maps that indicate the extent of each species on a global scale.

Ashby, Maurice. *Introduction to Plant Ecology*. New York: St. Martin's Press, 1961.
Ashby, an English scientist, describes in this book the ecology and distribution of plant species based upon influencing environmental factors. Chapter 6, "Biotic Factors in the Environment," looks at factors such as temperature, humidity, and day-length and their influences upon the ability of various plant species to succeed. Chapter 7, "Shade Cast: Vegetative Reproduction and Spread," examines plant competition for available resources and plant reproductive capacities.

Barnes, Robert D. *Invertebrate Zoology*. Philadelphia: W. B. Saunders, 1968.
Barnes, a Gettysburg College professor, discusses the various species of soft-bodied invertebrates in this lengthy reference guide. Step-by-step, he proceeds through each major class of invertebrates, beginning with protozoa and sponges, then moving to coelenterates, flatworms, annelids (segmented worms), mollusks (such as snails or squid), arthropods (such as spiders), crustaceans (such as lobsters), insects, echinoderms (such as sea urchins), and so forth. Many detailed diagrams of invertebrate species accompany the text.

Beck, Charles S., ed. *Origin and Early Evolution of Angiosperms*. New York: Columbia University Press, 1976.

This extensive volume includes presentation papers from the first International Conference of Systematic and Evolutionary Biologists, held in Boulder, Colorado, in August, 1973. The evolution of flowering plants (angiosperms) over the past four hundred million years is presented in the numerous papers. N. F. Hughes describes "Cretaceous Paleobotanic Problems." Armen Takhtajan discusses "Neoteny and the Origins of Flowering Plants." G. Ledyard Stebbins examines "Seeds, Seedlings, and the Origin of Angiosperms." Other contributors to this valuable reference include James Doyle, Kenneth Sporne, Gilbert Brenner, Rudolf Schuster, and F. Ehrendorfer. Many photographs, maps, species identifications, and world species distributions are provided.

Boyce, Mark S., ed. *Evolution of Life Histories of Mammals: Theory and Pattern*. New Haven, Conn.: Yale University Press, 1988.

This book summarizes the evolutionary history of mammalian species, in so doing incorporating important topics and methodologies in evolution. Among the topics discussed include genetic variations within mammalian populations, natural selection for or against certain genes (traits) based upon environmental conditions, and the biogeographic distribution of mammalian species from an evolutionary basis. The population biology of mammalian species is described with the necessary mathematical rigor, although it is not too complicated.

Briggs, D., and S. M. Walters. *Plant Variation and Evolution*. New York: McGraw-Hill, 1969.

Briggs and Walters present a unique historical and scientific study of plant evolution for the layperson. The book begins with a presentation of English naturalist John Ray, who defined the concept of species. Darwin's botanical studies during his around-the-world voyage are described, along with the pioneering plant genetic work of Gregor Mendel and Hugo de Vries. Briggs and Walters describe mitosis and meiosis in plant reproduction, plant chromosomes, probable courses of plant evolution, and plant breeding (artificial selection) by humans. The book contains a very good bibliography.

Buchsbaum, Ralph. *Animals Without Backbones: An Introduction to the Invertebrates*. Chicago: University of Chicago Press, 1938, 1948, 1976.

Breathtaking photographs highlight this fascinating introduction to invertebrate species for the layperson. Buchsbaum follows the traditional style of presenting the invertebrates class-by-class. He begins with the single-celled eukaryotic protozoans, then moves to such multicellular forms as the poriferans (sponges), coelenterates (such as hydra or jellyfish), flatworms, and segmented worms (such as earthworms). He continues with more complicated forms such as arthropods, crustaceans, insects, and echinoderms. He distinguishes marine, freshwater, and land forms for each class.

Caras, Roger. *The Endless Migrations*. Illustrations by Kimio Honda. New York: E. P. Dutton, 1985.

Caras, an ABC News correspondent, presents a down-to-earth tale of various mammalian, avian, and insect species during their yearly migrations from region to region in this entertaining book. He looks at the migrations of whales, lemmings, ducks, hummingbirds, and monarch butterflies. He describes the preparations and difficulties encountered by these species during their various journeys.

Caufield, Catherine. *In the Rainforest: Report from a Strange, Beautiful, Imperiled World*. Chicago: University of Chicago Press, 1984.

Caufield's story is an intriguing exploration of nature in the Amazon rain forest of South America. In chapter 1, "A Town in Brazil," she describes her travel to the town of Tucurui. There, she gains an appreciation for the importance of the rain forest to life on earth, a fact that she stresses throughout the work. Chapters 5, "Living in the Forest," and 13, "Fruits of the Earth," demonstrate the uses of forest species for humanity and the senseless destruction of nature by humans.

Cheetham, Alan H., and Albert J. Rowell, eds. *Fossil Invertebrates*. Palo Alto, Calif.: Blackwell Scientific Publications, 1987.

This enormous volume contains numerous survey and research articles by leading invertebrate zoologists. Invertebrate species are described phylum-by-phylum and class-by-class. Contributors include James W. Valentine, who provides two summary articles: "Invertebrate Organization: A Review" and "Evolution." Kenneth M. Towe describes the techniques of "Fossil Preservation." Michael E. Taylor illustrates "Biostratigraphy and Paleobiogeography." Other contributing authors include J. Keith Rigby, Richard A. Robinson, James Sprinkle, and David Clark.

Colbert, Edwin H. *Evolution of the Vertebrates: A History of the Backboned Animals Through Time*. 1955. 3d ed. New York: John Wiley & Sons, 1980.

This interesting book describes the development of vertebrate species during the past four hundred million years, beginning with the jawless fishes and proceeding through the amphibians (the first land vertebrates) to mammals. Good fossil and species diagrams support the text. Chapter 4, "Success of the Fishes," discusses the evolutionary adaptations of cartilaginous fish (such as sharks or rays) and bony fish (such as trout, perch, and hundreds more). Chapter 5, "From Water to Land," analyzes the evolution of air-breathing fish into early amphibians. Later chapters are devoted to the evolution of cetaceans and primates.

Diole, Philippe. *The Errant Ark: Man's Relationship with Animals*. New York: G. P. Putnam's Sons, 1974.

Diole's book is a call for action to preserve the environment and to halt the senseless destruction of animal and plant species. Jacques Yves-Cousteau writes the foreword to this compelling book on environmental conservation. Diole addresses humanity's callous disregard for fellow animal species. He cites hunters and politicians alike in their manipulation of laws to continue the destruction of many species and their habitats.

Eisenberg, John F. *The Mammalian Radiations: An Analysis of Trends in Evolution, Adaptation, and Behavior*. Chicago: University of Chicago Press, 1981.
This lengthy, detailed work is a comprehensive study of the evolution of mammalian species during the past two hundred million years. Part 1, "The Historical Perspective," looks at the history of mammals, their presumed evolutionary phylogenies, prototheria and metatheria, and adaptive radiations of new mammalian species into previously unoccupied environmental niches. Part 3, "Macrophysiology and Adaptations," looks at physical characteristics that make mammals so successful. Part 4, "Behavior as a Subject for Study," examines the highly social behaviors of mammalian species.

Fritzsch, Bernd, et al., eds. *The Evolution of the Amphibian Auditory System*. New York: John Wiley & Sons, 1988.
This seven-hundred-page compendium of scientific papers analyzes an important event in the evolution of land vertebrates, the evolution of hearing in amphibians (such as toads, frogs, or salamanders). William E. Duellman derives phylogenetic relationships between amphibian species in chapter 2, "Evolutionary Relationships of the Amphibians." Edwin R. Lewis and R. Eric Lombard study "The Amphibian Inner Ear." Wolfgang Walkowiak describes "Central Temporal Coding" of auditory impulses within the brain. Other contributors include Hans Schneider, Catherine A. McCormick, and Murray J. Littlejohn.

Good, Ronald. *The Geography of the Flowering Plants*. New York: John Wiley & Sons, 1984.
Good provides an interesting survey of angiosperms, the flowering plants that encase their seeds within fruit, in this lengthy, but informative, book. He details the structure, life cycles, and major species of angiosperms. He describes their worldwide distributions, methods of seed dispersal (such as wind, water, and animals), and the probable evolutionary history of angiosperms during the past two hundred million years. Angiosperms dominate the plant life of the current Cenozoic Era.

Guthrie, R. Dale. *Frozen Fauna of the Mammoth Steppe: The Story of Blue Babe*. Chicago: University of Chicago Press, 1989.
Paleontologist Guthrie describes the evolutionary development of mammoths and other large, extinct mammals of several tens of thousands of years ago in this

summary of an Alaskan paleontological expedition. He nicknames a thirty-six-thousand-year-old fossilized bison "Blue Babe" due to its preservation in blue mineral crystals. He analyzes fossil data from the Alaskan find to construct a picture of mammalian life thousands of years ago on the grassy steppe of a warmer Alaska.

Hall, R. P. *Protozoology*. Englewood Cliffs, N.J.: Prentice-Hall, 1953.
Hall's lengthy, but clear, book is an excellent introduction to microscopic, single-celled eukaryotes and constitutes an invaluable reference guide. He describes the major groups of protozoa, which are animal-like protists, and the many species within each group, including the mastigophorans, ciliophorans, and sporozoans. He details the anatomies, physiologies, genetics, and reproductive patterns of the major species within each group. Numerous, extensive diagrams highlight the text.

Hanson, Earl D. *The Origin and Early Evolution of Animals*. Middletown, Conn.: Wesleyan University Press, 1977.
Hanson's lengthy work, which employs a tremendous supply of reference citations, is a guide to the evolution of animal species on earth during the past five hundred million years. The work is progressive, moving through invertebrate animal species group-by-group from zooflagellates to amoebae to radiolarians to ciliates to sponge to cnidarians (such as jellyfish) to flatworms to turbellarians. Higher vertebrate animal species are excluded. Hanson describes the structures and evolution of these invertebrate species. He employs considerable scientific data, including DNA sequencing, to construct probable phylogenetic relationships between these species.

Holldobler, Bert, and Edward O. Wilson. *The Ants*. Cambridge, Mass.: Harvard University Press, 1990.
This large, beautifully illustrated volume is a complete analysis of the most abundant group of animal species on earth. It is written by two of the top world authorities on insects. Holldobler and Wilson describe the characteristic anatomies, physiologies, and behaviors of each of nearly three hundred ant genera. The evolution of ant species is discussed, including competition between individuals of the same and different species as well as natural selection for physical traits and social structure. A magnificent work.

Hutchinson, J. *Evolution and Phylogeny of Flowering Plants: Dicotyledons, Fact and Theory*. New York: Academic Press, 1969.
Hutchinson, a botanist with the Royal Botanical Gardens in Kew, England, has assembled a large reference work that illustrates the distribution and evolution of all families, genera, and species that fall within the classification of dicotyledon angiospermae, flowering plants that have two seed leaves. Utilizing a clearly

written text and hundreds of drawings, Hutchinson describes the characteristics of many diverse plant species, their geographical distribution on earth, and their probable evolutionary phylogenies.

Huxley, Anthony. *Green Inheritance: The World Wildlife Fund Book of Plants*. Foreword by David Attenborough. Garden City, N.Y.: Anchor Press/Doubleday, 1985.
Huxley's beautiful book, which is loaded with spectacular photographs, is aimed at promoting awareness of the importance of plants for all life on earth, including applications of plant species to human health and welfare. Humans are rapidly deforesting the planet, a process in which tens of thousands of plant species, many of unknown value, may be destroyed forever. Chapter 7, "Green Medicine," demonstrates the medicinal uses of many plant species. Chapters 2, "Guardians of the Environment," and 11, "Saving the Plants That Save Us," emphasize the importance of protecting our fragile planet.

Little, Colin. *The Colonisation of Land: Origins and Adaptations of Terrestrial Animals*. New York: Cambridge University Press, 1983.
This detailed, clearly written book thoroughly describes the evolution of land animals from four hundred million years ago to the present. Little, a zoology lecturer at the University of Bristol, describes flatworms, annelids (segmented worms such as earthworms), mollusks (such as snails or octopus), crustacea (such as lobsters), arachnids (spiders), insects, and vertebrate (such as amphibians, reptiles, birds, or mammals) animal species in great detail. He provides a wealth of data, charts, and photographs to illustrate the characteristics of these species and their probable common evolutionary heritage.

___________. *The Terrestrial Invasion: An Ecophysiological Approach to the Origins of Land Animals*. New York: Cambridge University Press, 1990.
In this sequel to *The Colonisation of Land*, zoologist Little discusses and speculates upon the physical basis of animal land colonization four hundred million years ago. He compares terrestrial versus aquatic ecosystems and necessary species adaptations for each type of environment. The evidence for an animal move to land includes major changes in animal respiratory physiology, alterations in the structure of the heart, nitrogen excretion in waste products, changes in sensory organs, and altered reproductive cycles. He cites numerous experiments involving many different animal species. He includes a look at intermediate environments in the sea-to-land transition, including swamps, estuaries, rivers, and streams.

Lorenz, Konrad. *On Aggression*. Translated by Marjorie Kerr Wilson. New York: Harcourt, Brace & World, 1963.
Lorenz, a Nobel laureate in Physiology or Medicine for his pioneering animal

behavior research, describes one of the most pronounced animal behaviors in this classic book for a general audience. Lorenz describes the evolution of behavior, "what aggression is good for," and how aggressive behavior has enhanced the evolution of various species. He concludes the book with an analysis of aggressive behavior in humans. His writing is lucid and witty, as usual, in this definitive study of animal behavior.

McLoughlin, John C. *Synapsida: A New Look into the Origins of Mammals*. New York: Viking Press, 1980.
In this thought-provoking work, McLoughlin approaches mammalian evolution from a different angle by proposing that all mammals are descended from reptilian synapsids. He supports his arguments with a large body of scientific data, including comparisons of structural changes in joints of mammals and ancient reptiles. "The Making of the Mammals," a later chapter, analyzes the evolution of cetaceans (such as whales, dolphins, and seals) and primates.

Maglio, Vincent J., and H. B. S. Cooke, eds. *Evolution of African Mammals*. Cambridge, Mass.: Harvard University Press, 1978.
This extensive data reference on African mammals past and present includes a lengthy text and numerous diagrams, photographs, and fossil sketches. In chapter 1, R. C. Bigalke looks at the "Present-Day Mammals of Africa." In chapter 3, A. W. Crompton and F. A. Jenkins, Jr., describe our mammalian ancestors in "Mesozoic Mammals." Alan C. Walker examines "Prosimian Primates" in chapter 6, and "Cenozoic Apes" are covered in chapter 8 by E. L. Simons, Peter Andrews, and David R. Pilbeam.

Manning, Aubrey. *An Introduction to Animal Behavior*. Reading, Mass.: Addison-Wesley, 1979.
Manning's work is a simple, yet very comprehensive, introduction to the study of ethology (animal behavior). Chapters 1, "Reflexes and Complex Behavior," and 3, "Stimuli and Communication," deal with basic processes used by animals to communicate. Chapter 5 deals with aggression and competition between individuals. Chapter 6, "Evolution," includes an extensive section focusing upon "The Adaptiveness of Behavior." Chapter 8, "Social Organization," looks at sociobiology, social insect species (such as ants, bees, or termites), and primates.

Margulis, Lynn, and Dorion Sagan. *Origins of Sex*. New Haven, Conn.: Yale University Press, 1986.
Evolutionary biologist Margulis and writer Sagan combine to produce a work illustrating how a single phenomenon, sex, has evolved over the past one billion years. They trace the origins of sexual reproduction in prokaryotic (bacterial) cells to its elaborate development in multicellular eukaryotes such as ourselves. They illustrate the basic processes of evolution, including mutation, natural selection,

and adaptive radiation of species. The emphasis is placed upon how evolution works at the cellular level, which is well-illustrated in the evolution of meiosis in reproductive cells.

Masterton, R. B., M. E. Bitterman, C. B. G. Campbell, and Nicholas Hotton. *Evolution of Brains and Behavior in Vertebrates*. Hillsdale, N.J.: Lawrence Erlbaum, 1976.
This five-hundred-page volume includes scientific papers describing vertebrate evolution and intelligence by leading experts in the field. Mark A. Berkeley discusses visual acuity in various vertebrate species. W. C. Stebbins describes vertebrate auditory (hearing) apparati. Sven Ebbesson intricately illustrates the neurology of anamniotic vertebrates, those vertebrate species that lack a protective egg sac for developing embryos. Wally Welker considers brain evolution in mammalian species, while C. B. G. Campbell concentrates upon brain evolution in primates. William A. Mason talks about primate social behaviors.

Murray, John W., ed. *Atlas of Invertebrate Macrofossils*. New York: John Wiley & Sons for the Paleontological Association, 1985.
This superb reference on invertebrate paleontology and evolution proceeds through the various invertebrate species step-by-step via the established Linnaen classification scheme for organisms. Murray describes thousands of species each of sponge, cnidaria (such as hydra or jellyfish), mollusks (such as octopuses, squid, or snails), echinoderms (such as sea stars or sea urchins), and arthropods (such as spiders). Many tables, diagrams, and photographs enhance the readings.

O'Connor, Raymond J. *The Growth and Development of Birds*. New York: John Wiley & Sons, 1984.
O'Connor describes with great clarity the evolution of birds, their anatomical structures and physiologies, and their behavior. He incorporates a huge amount of research data and summary tables to illustrate key points. Chapter 1, "Patterns of Avian Development," explores the physiological development of birds from birth to death. Chapter 8, "Maintenance of Body Temperature," represents an important study of bird blood thermoregulation, a phenomenon widely seen elsewhere only in mammalian species. Chapters 10, "Imprinting," and 13, "Instinct and Learning," describe important aspects of avian social behaviors.

Orians, Gordon H., and Otto T. Solbrig, eds. *Convergent Evolution in Warm Deserts: An Examination of Strategies and Patterns in Deserts of Argentina and the United States*. Stroudsburg, Pa.: Dowden, Hutchinson & Ross, 1977.
Orians and Solbrig's work is a collection of research papers written by themselves and their immediate colleagues as part of the U.S. contribution to the fifty-seven-nation International Biological Program from 1965 to 1975. The goal of this international study was to assess the affects of changing climate upon species

survival and distribution worldwide. In chapter 1, the editors summarize "An Evolutionary Approach to Ecosystems." Solbrig and colleagues analyze "The Strategies and Community Patterns of Desert Plants" in chapter 4. Michael A. Mores and colleagues discuss "The Strategies and Community Patterns of Desert Animals" in chapter 5. Conservation of nature is stressed throughout these articles. Species interactions are described in terms of mathematical models.

Peattie, Donald Culross. *A Natural History of Trees of Eastern and Central North America*. Illustrations by Paul Landacre. Boston: Houghton Mifflin, 1948, 1949, 1950.

In this lengthy survey of flowering and nonflowering trees, Peattie presents the various species genus-by-genus and species-by-species. He describes the nomenclature of trees and the geographical distributions of major tree species. Each species is described, including diagrams of leaf and bark structures. The sketches in this book are tremendous. Peattie's book is an invaluable guide for the serious botanist and student of angiosperm/gymnosperm evolution.

Pennak, Robert W. *Fresh-Water Invertebrates of the United States*. New York: John Wiley & Sons, 1978.

Pennak, a University of California biologist, discusses the various groups of invertebrate species, whether marine, freshwater, or terrestrial, in this eight-hundred-page guide for the beginning student. He describes invertebrates ranging from protozoa to sponge (more than eight thousand species) to mollusks to insects to echinoderms. He describes the characteristics of each species, including its development and its enemies. Excellent diagrams supplement the text.

Ridley, Mark. *The Explanation of Organic Diversity: The Comparative Method and Adaptations for Mating*. Oxford, England: Clarendon Press, 1983.

In this interesting book that deals with the genetic and behavioral aspects of reproductive behavior in vertebrate animal species, especially mammals, Ridley investigates two principal phenomena: (1) precopulatory mate guarding, and (2) homogamy, or assortative mating. He speculates about the probable biological bases for these two mating behaviors. He cites numerous species examples, many important research studies, and models describing mating behavior.

Romer, Alfred Sherwood. *The Vertebrate Body*. 4th ed. Philadelphia: W. B. Saunders, 1970.

Romer's book is a detailed, but clear, study of the anatomy and physiology of vertebrate animal species, concentrating upon specific characteristics such as the evolution of a chambered heart, the development of a highly complex nervous system, thermoregulation, two paired appendages, and a skeleton composed principally of bone and/or cartilage. Chapter 2, "The Vertebrate Pedigree," describes various vertebrate species beginning with acorn worms. Chapter 5 describes "The Early Development of Vertebrates" more than four hundred million years ago. Individual chapters are devoted to each of the many vertebrate organ systems (such as skeleton, muscles, circulation, nerves, and the endocrine system).

Savage, R. J. G., and M. R. Long. *Mammal Evolution: An Illustrated Guide*. New York: Facts on File, 1986.
This informative book, written by Savage and illustrated by Long, traces the evolution of mammalian species during their two-hundred-million-year history and charts their worldwide distributions today. Chapter 5 examines "Insectivores," mammals that eat insects. Chapter 6, "Carnivores," explores the lives of meat-eating mammals (such as cats or dogs). Chapter 7, "Paddlers and Swimmers," looks at the evolution of dolphins, whales, seals, and so forth, over the past seventy million years. Chapter 12, "Hoofed Herbivores," studies grazing mammals such as cows, horses, and wildebeest, while chapter 13, "Of Men and Monkeys," examines the primates.

Seeley, Thomas D. *Honeybee Ecology: A Study of Adaptation in Social Life*. Princeton, N.J.: Princeton University Press, 1985.
Seeley presents evolution from the analysis of a single species: the honeybee. He studies the evolution of honeybees, especially with respect to their behavior and communication. He also looks at the natural selection between individuals and between colonies (hives) via competition. Seeley's book is concise and well illustrated, and it sheds new light upon the evolution and ecology of honeybees and related social insects.

Shields, William M. *Philopatry, Inbreeding, and the Evolution of Sex*. Albany: State University of New York Press, 1982.
This unusual theoretical treatment of the evolution of sexual reproduction looks at mating mechanisms in a variety of species populations. The author discusses population genetics, inbreeding, and philopatry. Philopatry is a phenomenon in many species where an individual organism remains near the site of its birth and reproduces there; such a situation promotes inbreeding in the population, a condition that should be disadvantageous to the evolution of the population. The author Shields describes "The Disadvantages of Inbreeding" in chapter 4 and "Low Fecundity Species: Population Structure, Inbreeding, and Evolution" in chapter 5.

Silcock, Lisa, ed. *The Rainforests: A Celebration*. Compiled by the Living Earth Foundation. New York: Chronicle Books, 1989.
This glorious book is a collection of hundreds of photographs illustrating the incredible diversity of valuable animal and plant species that inhabit the threatened rain forests of the earth's tropical regions, rain forests essential to the stability

and maintenance of all living ecosystems on earth. The book, which includes a foreword by England's Prince Charles, is a project by the English educational Living Earth Foundation. Proceeds from the book support environmental education for children in Cameroon.

Sturkie, P. D., ed. *Avian Physiology*. New York: Springer-Verlag, 1976. 4th ed. 1986.
Sturkie's work is a chapter-by-chapter look at each of the principal body organ systems within the bird body, in each case analyzing the anatomical structures of importance and the physiological mechanisms by which the entire organism exists. In chapter 1, "The Nervous System," C. A. Benzo describes the evolution and functioning of a highly developed brain as compared to those of reptilian ancestors. In chapter 8, "Respiration," M. R. Fedde describes the development of a very efficient four-chambered heart and accompanying lungs in birds. G. C. Whittow discusses the importance of "Energy Metabolism" in chapter 10. Other chapters cover such things as the endocrine and immune systems.

Tiffney, Bruce. *Geological Factors and the Evolution of Plants*. New Haven, Conn.: Yale University Press, 1985.
Bruce Tiffney, D. J. Chapman, S. F. Barrett, R. Beerbower, and James W. Valentine are among fourteen leading evolutionary biologists who have contributed to this volume describing the evolution of plant species during the past one billion years. They cite geological evidence to construct scenarios for the evolutionary development of plants in various portions of the world and to surmise probable climatary conditions at various times during earth's history. The role of special evolutionary plant phenomena such as continental ecosystems and wildfire are discussed in several of these papers.

Torrey, Theodore W. *Morphogenesis of the Vertebrates*. New York: John Wiley & Sons, 1962, 1967.
Torrey describes the major physical features and evolutionary adaptations in the various classes of vertebrate species, including fish, amphibians, reptiles, birds, and mammals. He describes the comparative embryological study of various evolutionarily related species in chapter 1, "History of the Human Body." Subsequent chapters deal with a detailed examination of vertebrate evolution over the last four hundred million years and with studies of individual vertebrate organ systems, including the nervous, circulatory, reproductive, and digestive systems.

Turner, Bruce J., ed. *Evolutionary Genetics of Fishes*. New York: Plenum Press, 1984.
The evolution of fish species is described in this thorough survey of the subject. The articles in this book deal with fish evolution from a molecular genetic approach. Fred G. Allendorf and Gary H. Thorgaard discuss the development

of "Tetraploidy and the Evolution of Salmonid Fishes." Donald C. Morizot and Michael J. Sicilians describe "Gene Mapping in Fishes and Other Vertebrates." Irv Kornfield studies the "Descriptive Genetics of Cichlid Fishes." The papers are loaded with relevant molecular data that demonstrate species relatedness.

Walker, Ernest P., et al. *Mammals of the World.* Vols. 1-3. Baltimore: The Johns Hopkins University Press, 1964.
This detailed work is a family-by-family description of all mammalian species, highly social and intelligent animals that feed their young with milk and that generally have body hair/fur. The characteristics of each species is described, along with anatomies, life cycles, and photographs. The worldwide distribution of each mammalian genera is presented. Numerous references are cited for further study in this important reference guide.

Wessells, Norman K. Introduction to *Vertebrate Adaptations: Readings from Scientific American.* San Francisco: W. H. Freeman, 1952-1968.
This important reference work is a collection of *Scientific American* reprint survey articles that constitute authoritative presentations of vertebrate evolutionary adaptations. Milton Hildebrand describes "How Animals Run," John H. Storer examines "Bird Aerodynamics," Carl J. Wiggers looks at "The Heart," and Julius Comroe, Jr., studies "The Lung," all features that are unique to vertebrate animal species. Also, Charles M. Bogert explains "How Reptiles Regulate Their Body Temperature," and W. H. Thorpe studies "The Language of Birds." The articles are beautifully illustrated and written specifically for the layperson.

Wilkins, Adam S. *Genetic Analysis of Animal Development.* Illustrations by Belinda Dumant. New York: John Wiley & Sons, 1986.
Wilkins' detailed, but clear, presentation of animal development looks at three species in particular: the nematode *Caenorhabditis elegans*, the fruit fly *Drosophila melanogaster*, and the mouse *Mus musculus*. Embryological development in each of these species is described from the oocyte stage to the fertilization by sperm to the blastoderm stage. Blastoderm is followed by gastrulation and development towards maturity. Cell lineages and genetics are analyzed, and applications of cloning to genetic studies of animal development are discussed.

Woodwell, G. M., ed. *The Role of Terrestrial Vegetation in the Global Carbon Cycle: Measurement by Remote Sensing.* New York: John Wiley & Sons, 1984.
This important volume, compiled in collaboration with the Scientific Committee on Problems of the Environment, is a collection of reports, maps, and photographs that demonstrate the tremendous importance of land and ocean plant life in the proper recycling of carbon and other life materials throughout the environment. The reports herein emphasize the dangers of deforestation and related insults to the environment that have been propagated by humans during the

twentieth century. Contributors include Woodwell of the Marine Biological Laboratory Ecosystem Center at Woods Hole, Massachusetts, F. C. Billingsley of the NASA Jet Propulsion Laboratory in Pasadena, California, and W. H. Schlesinger of Duke University.

Chapter 9
DEBATES OVER EVOLUTIONARY DESCENT

Phylogenetic Relationships Between Species

Abbott, Lois A., Frank A. Bisby, and David J. Rogers. *Taxonomic Analysis in Biology: Computers, Models, and Databases*. New York: Columbia University Press, 1985.
Abbott, Bisby, and Rogers demonstrate the importance of computer simulations and analyses of evolutionary evidence (such as comparative DNA and protein sequences between various species, fossil comparisons, and comparative anatomy between species) to construct phylogenetic trees relating different species and to outline probable evolutionary histories of major groups of species. They describe older methods of phylogenetic reconstruction and then introduce the advantages of computer analysis. They present models, algorithms, character analysis, and database management in their discussions of experimental methodologies.

Agassiz, Louis. *Methods of Study in Natural History*. Boston: Ticknor & Fields, 1863. Reprint. New York: Arno/The New York Times, 1970.
In this dated, but extremely accurate, work, Agassiz, a contemporary of Charles Darwin, presents an early account of established methods for classifying organisms, including categories of species classification in chapter 3. In chapter 5, "Different Views Respecting Orders," he examines alternative and contradictory viewpoints concerning the probable relatedness between some species. In later chapters, he looks at differences between species and breeds (chapter 10), the formation of coral reefs (chapter 11), and homologies between various species (chapter 13).

Alexander, R. McNeill. *Dynamics of Dinosaurs and Other Extinct Giants*. New York: Columbia University Press, 1989.
Alexander's work combines an outstanding text, good sketches, summary data charts, and skeletal reconstructions of dinosaurs in a clear discussion of dinosaur origins, history, extinction, and relatedness to current living species. Chapter 7, "Hot-Blooded Dinosaurs," examines Robert T. Bakker's views that dinosaurs may have been warm-blooded, unlike other reptiles. Chapter 11, "Giant Birds," looks at *Archaeopteryx* and other flying reptiles as possible ancestors of birds.

Anderson, D. T. *Embryology and Phylogeny in Annelids and Arthropods*. New York: Pergamon Press, 1973.
In this important work, Anderson examines the embryological evidences for evolution in some of the most prevalent invertebrate animal species on earth, namely the annelids (segmented worms such as earthworms), arthropods (such

as spiders and scorpions), insects (numbering at more than one million species), and crustaceans (such as lobsters and crayfish). Anderson links these species and suggests possible phylogenies for their evolutionary descent over the past five hundred million years. Each major group of species is well represented and illustrated.

Anderson, O. Roger. *Comparative Protozoology: Ecology, Physiology, Life History.* New York: Springer-Verlag, 1988.
Anderson's fine work is an entertaining analysis of the abundant microscopic protozoa, animal-like species that very probably are the ancestors of all animals. He relies upon a wealth of data to arrive at possible phylogenetic relationships between various groups of protozoa and animals during evolutionary time. He examines species of each major protozoan phylum and subphylum, including the flagellates, the amoebae, and the ciliates. Principal physiological adaptations and nutritional requirements of each species group are explained in detail.

Andrews, Henry N., Jr. *Studies in Paleobotany*. New York: John Wiley & Sons, 1961.
Andrews, a botanist at Washington University in St. Louis, presents a unique paleontological look at fossil plants and their evolution either into modern plants or into extinction. He provides a class-by-class analysis of fossilized plants. He describes the evolution of ferns, gymnosperms, and angiosperms and the probable phylogenetic relationships between various living species. He cites points of dispute over possible phylogenies. He provides chapters on methods for isolating and interpreting fossil plant remains, the importance of seed evolution in plant evolution, and fossil plants of various continents, including the Arctic and Antarctic regions. Charles J. Felix of Sun Oil Company presents a chapter describing polynology.

Brooks, Daniel R., and Deborah A. McLennan. *Phylogeny, Ecology, and Behavior: A Research Program in Comparative Biology*. Chicago: University of Chicago Press, 1991.
Evolutionary biologists Brooks and McLennan present a new look at how species may evolve in this interesting, authoritative work. Instead of analyzing a species' evolution based upon its phylogenetic relationship to other species, they examine the origins of new species and interactions between species within complex ecosystems. They stress the point that evolution is more of a process in which communities of species within single ecosystems coevolve over time and give rise to new species.

Claugher, D., ed. *Scanning Electron Microscopy in Taxonomy and Functional Morphology*. Oxford, England: Clarendon Press for the Systematics Association, 1990.

Several electron microscopists contribute papers to this volume illustrating the use of powerful scanning magnification to discern the possible evolutionary relatedness between various species. Scanning electron microscopy can produce three-dimensional views of objects magnified up to one hundred thousand times. Fantastic scanning electron photomicrographs illustrate similarities and differences between numerous species both fossilized and living, particularly plant species. M. Wagener, C. G. Ogden, and W. Barthlott are among the contributors to this interesting work.

Clifford, H. T., and W. Stephenson. *An Introduction to Numerical Classification.* New York: Academic Press, 1975.
Clifford and Stephenson's work is a theoretical analysis of the methods used to classify species and to determine the possible evolutionary phylogenies relating different groups of species. They examine important, yet subtle, topics such as the identification of important morphological structures on organisms, the wide diversity of traits between individuals of a species and between species, and proposed mathematical models for describing the process of speciation. Their work is a strong introduction to the subject for the student of biology and evolution.

Cole, A. J., ed. *Comparative Taxonomy.* Proceedings of the Colloquium in Numerical Taxonomy held at the University of St. Andrews, September, 1968. New York: Academic Press, 1969.
This book is a very mathematical treatment of species classification and the construction of phylogenetic trees based upon fossil and molecular data from different species. Cole, A. S. Boughey, H. D. Abramson, R. L. Middleton, and Charles C. Ostrander are among the conference participants who present lively discussions of problems with phylogenetic reconstructions. Abramson demonstrates the uses of computers in studying evolutionary evidence.

Crowson, R. A. *Classification and Biology.* New York: Atherton, 1970.
In this very clear work, Crowson introduces the beginning student to the methods by which animal and plant species are classified. Crowson demonstrates the Linnaen classification scheme of kingdom, phlum, class, order, family, genus, species, subspecies, and so forth with a number of species examples. Crowson introduces the reader to the nomenclature of species classification, the fossil record as applied to classification of species, and the use of chromosomes and molecular genetics to classify and relate different species. Chapter 22, "Epilogue: The Future of Systematics," speculates upon coming trends in the study of organismal evolution. Many references are cited in this work.

Daber, Rudolf, and Jochen Helms. *Fossils: The Oldest Treasures That Ever Lived.* Neptune City, N.J.: T.F.H., 1985.

In this book, Daber and Helms examine various lines of evolutionary descent based upon the fossil paleontological record. They consider various important events in animal and plant evolution and articles of dispute concerning the evolutionary relationships between various species. The possible ancestor of modern birds, the giant flying reptile *Archaeopteryx*, is examined in chapters entitled "Archaeopteryx, the Oldest Known Bird" and "Confusion over the Feathers of Archaeopteryx." A later chapter, "Great Ideas of the Laws of Development," examines various aspects of fossil plant evolution.

Delevoryas, Theodore. *Biology Studies: Morphology and Evolution of Fossil Plants*. New York: Holt, Rinehart and Winston, 1962.
Delevoryas, a Yale University scientist, presents the science of paleobotany in this short book. In chapter 1, "Introduction," he discusses various physical mechanisms by which ancient species became preserved in the soil or rock formations. He then proceeds through a series of many chapters in which he summarizes the evolution and taxonomy of the various divisions and subdivisions of nonvascular and vascular plants. He discusses possible phylogenies for these plants and alternate views on fossil plant interpretations.

Duncan, Thomas, and Tod F. Stressy, eds. *Cladistics: Perspectives in the Reconstruction of Evolutionary History*. New York: Columbia University Press, 1984.
Duncan and Stressy's work is a compilation of scientific papers presented at the 1981 Workshop on the Theory and Applications of Cladistic Methodology held at the University of California, Berkeley. The participants at this workshop tackled major problems in the classification of species and the reconstruction of probable evolutionary events during the past few hundred million years. The papers presented herein describe cladistic theory and the increasing use of computers to model evolution and speciation. Contributors include Christopher Meacham, Joseph Felsenstein, Daniel R. Brooks, Gareth Nelson, and David Hull.

Eldredge, Niles, and Steven M. Stanley, eds. *Living Fossils*. New York: Springer-Verlag, 1984.
Eldredge and Stanley have assembled a collection of essays that deal with the evolution of plants and animals, the fossil record, and debates concerning the evolutionary relatedness between various species. Several articles are devoted to arrested evolution, a process in which there has been little change between a fossil species of many millions of years ago and current members of the same species. Such species are termed "living fossils." Eugene R. Meye and Peter Forey consider crocodiles and coelocanths, respectively, as such living fossils. Other contributors include Ian Tattersall and Jeffrey H. Schwartz.

Felsenstein, J., ed. *Numerical Taxonomy*. New York: Springer-Verlag, 1983.
This lengthy collection of theoretical papers resulted from a July, 1982, NATO

Institute on Numerical Taxonomy held at Bad Windsheim, Germany. Many leading evolutionary scientists tackle problems in the interpretation of the fossil and genetic records of life on earth, competing theories on species evolution and relatedness, mathematical modeling of evolution, and the applications of computers to species classification. Contributors include Joel Cracraft, Hai K. Seth, F. James Rohlf, and J. Felsenstein.

Fenton, Carroll Lane, and Mildred Adams Fenton. *The Fossil Book: A Record of Prehistoric Life*. 1958. Revised and expanded by Patricia Vickers Rich, Thomas Hewitt Rich, and Mildred Adams Fenton. New York: Doubleday, 1989.
This beautiful, 740-page masterpiece contains a captivating text and hundreds of photographs, sketches, and illustrations. The Fentons describe fossils and their formation in chapter 2, "Rocks, Fossils, and Ages." They discuss various viewpoints on species relatedness and taxonomy in chapter 4, "Groups, Names, and Relationships." Chapters 6 through 30 deal with the successive evolution of eukaryotic species, beginning with the single-celled protists and building up to the dinosaurs and the possible dinosaur descendants, the birds.

Ferguson, Andrew. *Biochemical Systematics and Evolution*. New York: John Wiley & Sons, 1980.
In this short work, Ferguson presents the classification of organisms and the construction of evolutionary phylogenies that illustrate the probable evolutionary relationships between different species by using biochemical data and mathematical models. He examines the molecular basis of evolution by utilizing comparative DNA, chromosome, and protein studies between different species (such as comparisons between chimpanzees and humans). Increasing differences in molecular codes between various species indicates distant relatedness in evolutionary time. Few differences in genetic codes between species indicates close relatedness and common origins.

Funk, V. A., and D. R. Brooks, eds. *Advances in Cladistics: Proceedings of the First Meeting of the Willi Hennig Society*. Bronx, N.Y.: New York Botanical Garden, 1981.
Funk, Brooks, J. S. Farris, and M. R. Bolick are among the dozens of contributors to this volume discussing advances in the classification of species, both extinct and extant (living), a science that goes by names such as cladistics, taxonomy, and systematics. The authors of the papers describe the utilization of molecular biological data (comparative DNA and protein sequences between species), biogeographical analyses of fossil distributions, and novel mathematical models to illustrate the processes of evolutionary change during the past several hundred million years.

Hallam, A., ed. *Atlas of Paleobiogeography*. New York: Elsevier Scientific, 1973.
This handsome book contains collected scientific papers that describe the fossil remains of many extinct species, their systematic and taxonomical classifications, their possible evolutionary relatedness to other species both extinct and extant, and their possible fates. Among the many contributors to the volume include Allison R. Palmer and H. B. Whittington, who each describe the fossil trilobites in independent papers. D. Edwards discusses "Devonian Floras." A. L. Panchen and C. Barry Cox both discuss tetrapods, albeit in different geological periods. Alan J. Charig contributes an interesting article on "Jurassic and Cretaceous Dinosaurs."

Hawksworth, D. L., ed. *Prospects in Systematics*. Systematics Association Special Vol. 36. Oxford, England: Clarendon Press, 1988.
A July, 1987, meeting of 140 paleobotanists at the Royal Society of London produced this informative collection of discussion papers aimed at solving problems in species classification and evolutionary theory. In paper one, "Systematics: The Keystone of Biology," D. L. Hawksworth and F. A. Bisby discuss the central role of species classification in the structure of all biological disciplines and in evolutionary theory. In paper seven, "The Contributions of Paleontology to Systematics and Evolution," A. Hollam illustrates how the fossil record has been central to the classification and evolutionary analysis of species. K. A. Joysey examines molecular biology and systematics.

Hennig, Willi. *Phylogenetic Systematics*. Translated by D. Dwight Davis and Rainer Zangerl. Urbana: University of Illinois Press, 1966.
In this comprehensive book, Hennig describes all angles of, and aspects to, the science of systematics, the identification and evolutionary classification of species. In chapter 1, Hennig introduces "The Position of Systematics Among the Biological Sciences." The problems of what constitutes a species and comparative morphology between species are tackled in chapter 2, "Tasks and Methods of Taxonomy." Chapter 3, "Problems, Tasks, and Methods of Phylogenetics," considers the problems of explosive radiations of new species, monophyly versus polyphyly, and species dichotomies.

Hoenigswald, Henry M., and Linda F. Wiener, eds. *Biological Metaphor and Cladistic Classification: An Interdisciplinary Perspective*. Philadelphia: University of Pennsylvania Press, 1987.
This interesting work consists of theoretical papers by leading scientists that examines problems in the way species are approached and scrutinized by systematists and taxonomists. The contributors, who represent a wide variety of scientific viewpoints, make suggestions for improving our views of species, their evolution, and their classification. H. Don Cameron, William S. Y. Wang, Anna Morpurgo Davies, E. F. Konrad Koerner, and Michael Novacek produce fascinating papers in this volume.

Humphries, Christopher J., and Lynne R. Parenti. *Cladistic Biogeography*. Oxford, England: Clarendon Press, 1986.
Humphries and Parenti, in a very short book, describe the evolutionary phylogenetic reconstruction of species relatedness and their worldwide distributions in this informative reference work. They describe the biogeography of species, the distribution and worldwide spread of various species over geological time. They relate biogeographical data to the construction of cladograms, phylogenetic trees that attempt to illustrate major evolutionary processes.

Huxley, Julian. *The New Systematics*. London: The Systematics Association, 1971.
This reprint of Huxley's lengthy, but brilliant, 1940 treatise on the systematic classification of organisms foresees the current uses of biochemical genetics to illustrate the relatedness between species and the evolutionary process. Huxley concentrates upon certain problems that even today plague the identification of species' origins, the biogeographical evolutionary distributions of species, and the fossil record. He cites the genetic Synthetic Theory of Evolution, which unites genetics and systematics, as proposed by Theodosius Dobzhansky. The synthetic theory is one of the major achievements of twentieth-century evolutionary biology.

Jaeger, Edmund C. *A Source-Book of Biological Names and Terms*. Springfield, Ill.: Charles C Thomas, 1944.
Jaeger's work is a guide to the serious systematist or taxonomist for the classification of new species. The book describes the methods by which species are classified. It also lists hundreds of prefixes and suffixes with definitions for the appropriate labeling and characterization of sample organisms. While the book is aimed at the serious student of evolutionary biology, it is easily accessible to the beginner.

Leone, Charles A., ed. *Taxonomic Biochemistry and Serology*. New York: Ronald Press, 1964.
Seven hundred pages of provocative, but clear, scientific articles highlight this important early work on molecular biology and taxonomy. Expert scientists such as Ernst Mayr, Robert Sokal, and Alan A. Boyden examine the science of systematics, which in 1964 was shifting to molecular analyses of DNA and protein data between different species. Konrad Bloch, Alan C. Wilson, and Morris Goodman present some pioneering papers describing the comparative biochemistry of species to illustrate evolutionary relationships. An excellent guide both for the research scientist and for the beginning student.

Margulis, Lynn, and Karlene V. Schwartz. *Five Kingdoms: An Illustrated Guide to the Phyla of Life on Earth*. Foreword by Stephen Jay Gould. San Francisco: W. H. Freeman, 1987.
Margulis and Schwartz's book is, for the beginning student, an excellent introduc-

tion to species classification and the problems of constructing accurate evolutionary phylogenies. They present a phylum-by-phylum array of species, in the process citing many data sources and references. All major groups of organisms on earth are described and illustrated. Great photographs and diagrams accompany the text.

Mayr, Ernst. *Principles of Systematic Zoology*. New York: McGraw-Hill, 1969.
This clear, simple book, written by one of the foremost evolutionary scientists of the twentieth century, discusses the collection of species and fossils, the interpretation of species and their relatedness, the construction of phylogenies, and problems of such phylogenetic reconstructions. In chapter 8, "The Qualitative and Quantitative Analysis of Variation," differences between species are assessed in terms of relating species in phylogenetic trees. Such problems are further elaborated in chapter 9, "Taxonomic Decisions on the Species Level."

____________, ed. *The Species Problem*. New York: Arno Press, 1974.
This fascinating collection of papers originated from an American Association for the Advancement of Science symposium held in Atlanta, Georgia, in December, 1955. A stellar list of evolutionary biologists consider such provocative questions as, "Do species exist?" The characteristics of species and the problems of classifying them are described from varying scientific viewpoints. Contributing authors include Mayr, Verne Grant, T. M. Sonneborn, C. Ladd Prosser, John Imbrie, and Hampton L. Carson.

Mielke, H. W. *Patterns of Life: Biogeography of a Changing World*. Boston: Unwin Hyman, 1989.
Mielke's thorough, but not too lengthy, discussion of evolution is an excellent introduction to the natural history of life on earth. Mielke describes the various physical processes that have shaped the planet, including volcanism, erosion, and plate tectonics/continental drift. The development of billions of species over the past several billion years of earth history is examined in detail along with biogeographic distributions of species. Controversies and debates over species origins and extinctions are examined as well. Good diagrams and excellent photographs support the text.

Moore, Raymond C., Cecil G. Lalicker, and Alfred G. Fischer. *Invertebrate Fossils*. New York: McGraw-Hill, 1952.
Moore, Lalicker, and Fischer's 766-page compendium is an extremely detailed summary of invertebrate evolution and systematic relatedness based upon the paleontological fossil record. Hundreds of sketches of species and possible evolutionary relationships between various species are provided. The authors examine the comparative morphologies between extinct and extant (living) invertebrate species. The clearly describe the geological strata from which

successive stages of invertebrate evolution have been unearthed. Invertebrate species are presented phylum-by-phylum and class-by-class.

Nelson, Gareth, and Norman Platnick. *Systematics and Biogeography: Cladistics and Vicariance*. New York: Columbia University Press, 1981.
Nelson and Platnick's work is a lengthy, detailed discussion of the historical debates concerning the biogeographical interpretation of the evolutionary record, the interpretation of the fossil record, and the proper classification of species. The authors examine many competing theories and interpretations of the evolutionary process as well as numerous possible alternate phylogenies that attempt to illustrate probable evolutionary relationships between distinct species and groups of species. Excellent diagrams illustrate alternative viewpoints in this thought-provoking book.

Otte, Daniel, and John A. Endler, eds. *Speciation and Its Consequences*. Sunderland, Mass.: Sinauer Associates, 1989.
Otte and Endler have assembled an impressive collection of scientific papers that tackle some major problems in evolutionary biology, specifically the problems of identifying and classifying species, the factors that make a species distinct from other species, the interpretation of fossil and molecular data in order to relate species, and the construction of possible evolutionary phylogenies. In the first paper, "The Meaning of Species and Speciation: A Genetic Perspective," the molecular geneticist Alan R. Templeton provides an insightful overview into the problem of what constitutes a species. Other contributors include Gareth Nelson, Godfrey Hewitt, N. H. Barton, David Sloan Wilson, and Douglas Futuyma.

Pankhurst, R. J., ed. *Biological Identification with Computers*. New York: Academic Press for the Systematics Association, 1975.
This volume represents the proceedings of a September, 1973, King's College (University of Oxford) symposium on systematics and taxonomy. The primary focus of the conference is the utilization of computers to improve the identification and classification of species and the establishment of suitable evolutionary phylogenies that demonstrate speciation/species relatedness. S. M. Walters describes traditional methods in systematics and taxonomy. R. J. Pankhurst describes "Identification by Matching" using computer analyses of data. R. R. Aitchison and J. H. Beamon also contribute to this volume.

Platnick, Norman I., and V. A. Funk, eds. *Advances in Cladistics*. Vol. 2 of the *Proceedings of the Second Meeting of the Willi Hennig Society*. New York: Columbia University Press, 1983.
A number of prominent evolutionary systematists and taxonomists, including Platnick, Funk, P. M. Richardson, David Hull, E. R. Sober, and R. H. Brady, have contributed research and survey papers to this book, which describes

improvements in the methods of species analysis and classification. The included papers deal with matters of theory, mathematical models of speciation and evolution, examples of speciation, and the computer analysis of comparative biochemical data. The book is a good survey of cladogram constructions, the development of phylogenetic trees to relate different species.

Potter, Stephen, and Laurens Sargent. *Pedigree: The Origins of Words from Nature.* New York: Taplinger, 1974.
This rather interesting work, dedicated to James Maxwell McConnell Fisher, is a semantic analysis of the naming and classification of species. Potter and Sargent address the nomenclature of species systematics and taxonomy, in the process gaining insight into the scientific rationale behind the descriptions of organisms. Much of species classification seems to be based upon various English dialects. Potter and Sargent's work is an unusual contribution to the field of species classification.

Ridley, Mark. *Evolution and Classification: The Reformation of Cladism.* New York: Longman, 1986.
In this extraordinary work, which is very understandable for the average person, Ridley examines the techniques and experimental methodologies for classifying and relating species both before and after Darwin's proposal of evolution by natural selection. Ridley considers alternative viewpoints and phylogenies describing species evolution in this scholarly work. He examines major debates over evolutionary theory. He also presents theoretical models for the evolution and extinction of species during the earth's history.

Salthe, Stanley N. *Evolving Hierarchical Systems: Their Structure and Representation.* New York: Columbia University Press, 1985.
Salthe's incredible book is a brilliantly alternative look at the order, arrangement, and evolution of life on earth. Salthe concentrates upon patterns of structure and form found throughout all living organisms on earth to derive several possible models of organic evolution. He discusses the hierarchy of nature and the repetition of specific patterns and arrangements in various tissues, organs, and organisms in a variety of different species. Salthe's book is a refreshing alternative view to the usual static taxonomy of living things.

Schenk, Edward T., and John H. McMasters. *Procedure in Taxonomy.* Stanford, Calif.: Stanford University Press, 1936.
This seventy-two-page work is a very important description of classification methods for linking isolated species by descent. It sets the basic parameters, used internationally, by which species are grouped and classified. The identification of a new species is presented in chapter 4, "The Description of a New Species." Chapter 7 describes the "Storage of Type Material." An appendix includes a reprint of the "International Rules of Zoological Nomenclature."

Scott-Ram, N. R. *Transformed Cladistics, Taxonomy, and Evolution*. New York: Cambridge University Press, 1990.
Scott-Ram's book is a clear, philosophical treatment of alternative viewpoints and theories concerning the formation of new species, the classification of organisms, and the evolution of life on earth. While the points of contention between competing theories may seem trite, Scott-Ram succinctly illustrates the importance of each position. The book is a good treatment of the problems of evolutionary theory and is readily understandable to the general public.

Shrock, Robert R., and William H. Twenhofel. *Principles of Invertebrate Paleontology*. 2d ed. New York: McGraw-Hill, 1953.
Shrock and Twenhofel's compendious volume contains a detailed, but clear, text and hundreds of species photographs, characteristics, and worldwide distribution maps. Chapter 1, "Introduction," describes paleontology and the manipulation/interpretation of fossils with respect to evolutionary classifications of species. The various classes of invertebrate species are described in presumed evolutionary order, beginning with protozoa, poriferans (sponges), and coelenterates (such as hydra) and proceeding to arthropods (such as spiders) and echinoderms (such as sea urchins or sea stars). Evolutionary debates over relationships between various groups are mentioned.

Sibley, Charles G., chair. *Systematic Biology*. Proceedings of an International Conference. Washington, D.C.: National Academy of Sciences, 1969.
The National Research Council sponsored this 1967 conference on systematics held at the University of Michigan, Ann Arbor. Good discussions highlight the papers presented in this informative work. Ernst Mayr and Frans Stafleu describe the history of systematics and taxonomy. Warren H. Wagner, Jr., and A. R. Krukeberg summarize the systematic classification of plant species, while Robert K. Selander presents the systematics of animal species classification. Herbert C. Dessamer looks at the uses of molecular data in species classification.

Simpson, George Gaylord. *Fossils and the History of Life*. A *Scientific American* book. New York: W. H. Freeman, 1983.
Simpson's book, which is dedicated to his late colleagues and eminent scientists Theodosius Dobzhansky, Julian Huxley, Bryan Patterson, and Alfred Romer, is a solid introduction to the geological record, fossil identification and interpretations, and controversies over the linkage of various species by phylogenetic means. He presents "Fossils as Living Organisms" in chapter 2. He elaborates upon problems in fossil interpretations in chapters 5, "Extinction, Origination, and Replacement," and 6, "Rates and Patterns of Evolution."

____________. *Principles of Animal Taxonomy*. New York: Columbia University Press, 1961.
Simpson, in his usual clear style, presents a lively scientific discussion of systematics, taxonomy, and problems in the analysis of species relationships on the evolutionary scale, especially with reference to animals, both invertebrate and vertebrate. He introduces these concepts in chapter 1, "Systematics, Taxonomy, Classification, and Nomenclature." He discusses classification problems in chapters 3, "Taxonomic Evidence and Evolutionary Interpretation," and 5, "The Species and Lower Categories."

Sneath, Peter H. A., and Robert R. Sokal. *Numerical Taxonomy: The Principles and Practice of Numerical Classification*. San Francisco: W. H. Freeman, 1973.
Sneath and Sokal present the mathematics of species classification and phlogenetic tree construction in this long, but clear, introduction for the beginning biology student. They present the statistical foundation of numerical taxonomy theory and the uses of mathematical models to interpret fossil and biochemical data. Alternate phylogenies that attempt to demonstrate species relatedness are described as well.

Wiley, E. O. *Phylogenetics: The Theory and Practice of Phylogenetic Systematics*. New York: John Wiley & Sons, 1976.
Wiley, a leading scientist in the field of phylogenetic classification, introduces the subject in this very clear book. He describes the nature of species, the factors that determine a species' distinctiveness from other species, and the construction of phylogenies to illustrate the evolution of a species during geological history. He analyzes species classification methods, the evolution of species, and patterns of reproduction within different species. He also addresses the nomenclature and rules for classifying species.

Mass Extinctions

Donovan, Stephen K., ed. *Mass Extinctions: Processes and Evidence*. New York: Columbia University Press, 1989.
In this important collection of research papers, Donovan and other scientists examine the paleontological and geological evidences for the occurrence of massive species extinctions over geological time, and even periodic mass extinctions as well. Donovan begins by setting "Paleontological Criteria for the Recognition of Mass Extinctions." S. R. Westrop considers "Trilobite Mass Extinctions Near the Cambrian-Ordovician Boundary in North America." P. J. Benchley likewise analyzes "The Late Ordovician Extinction." A. D. Barnosky looks at "The Late Pleistocene Event as a Paradigm for Widespread Mammal Extinctions." Other contributors to this volume include C. J. Orth and M. D. Brasier.

Elliott, David K., ed. *Dynamics of Extinction*. New York: John Wiley & Sons, 1986.
This exciting collection of summary research papers for and against the theory of periodic mass extinctions of species over geological time sprang from an August, 1983, symposium at Northern Arizona University following Louis and Walter Alvarez' proposal that the Cretaceous extinction sixty-five million years ago may have been precipitated by a comet or asteroid collision with earth. Contributors to this fascinating volume include J. John Sepkoski, Jr., David M. Raup, Digby McLaren, Edwin H. Colbert, Daniel Simberloff, and David Jablonski. Topics include Cretaceous, Permian, and Mesozoic extinctions, the causes of extinctions, and human-induced extinctions in tropical rain forests.

Hoffman, Antoni. *Arguments on Evolution: A Paleontologist's Perspective*. New York: Oxford University Press, 1989.
In this interesting work, Hoffman defends the neo-Darwinian theory of evolution and attacks scientists who muddle over trivial disputes concerning minor aspects of the theory, often merely for peer recognition. In chapter 4, he presents "The Fossil Record as Data on Evolution." In chapter 7, he analyzes "Punctuated Equilibrium: A Multitude of Interpretations." He considers periodic mass extinctions of species over geological time and their possible causes, both terrestrial and extraterrestrial, in chapter 11.

Hsu, Kenneth J. *The Great Dying*. San Diego: Harcourt Brace Jovanovich, 1986.
In this outstanding work, Hsu looks at evolution and alternate mechanisms for its occurrence. In chapter 1, "The Question of Fitness," he questions the attributes of organisms that supposedly give some individuals advantages over others during periods of evolutionary change. In chapter 7, "Death from the Sky," he considers the Alvarez proposal that the Cretaceous extinction and accompanying death of the dinosaurs may have been caused by an asteroid or cometary impact on earth. He also cites the 1908 Tunguska event in Siberia, where a small cometary piece probably collided with earth. In chapter 13, "The Race Is Not to the Swift," he reiterates this catastrophic scenario and emphasizes the element of chance in the evolution of life on earth.

Larwood, G. P., ed. *Extinction and Survival in the Fossil Record*. Systematics Association Special Vol. 34. Oxford, England: Clarendon Press, 1988.
This fascinating collection of scientific research papers focuses upon the questions of mass extinctions, the causes of such extinctions, and the possibility of extinctions occurring periodically. M. C. Boulter, R. A. Spicer, and B. A. Thomas examine "Patterns of Plant Extinction from Some Paleobotanical Evidence." M. J. Benton explores "Mass Extinctions in the Fossil Record of Reptiles: Paraphyly, Patchiness, and Periodicity(?)." D. M. Unwin looks at "Extinction and Survival in Birds," while R. J. G. Savage studies "Extinction and the Fossil Mammal Record." Other contributors include Colin T. Scrutton and Michael R. House.

Martin, Paul S., and Richard G. Klein, eds. *Quaternary Extinctions: A Prehistoric Revolution*. Tucson: University of Arizona Press, 1984.
Martin and Klein have assembled a collection of important research papers devoted to the occurrence and explanation of minor mammalian extinctions during recent geological history. S. David Webb presents "Ten Million Years of Mammal Extinctions in North America." N. K. Vereshchagin and G. F. Baryshnikov explore "Quaternary Mammalian Extinctions in Northern Eurasia." Richard G. Klein discusses "Mammalian Extinctions and Stone Age People in Africa." Other contributors to this thirty-eight-article, nine-hundred-page work include David R. Horton, Richard A. Kiltie, Elaine Anderson, and Larry D. Agenbroad.

Nitecki, Matthew H., ed. *Extinctions*. Chicago: University of Chicago Press, 1984.
Evolutionary scientist Nitecki has assembled an exciting collection of essays and scientific papers concerning the occurrence of species extinctions and possible periodicities in the occurrence of extinctions over geological history. David M. Raup presents "Death of Species." Alan Walker discusses "Extinction in Hominid Evolution" with reference to early human species and their fates. Andrew H. Knoll describes "Patterns of Extinction in the Fossil Record of Vascular Plants." Jared M. Diamond examines "'Normal' Extinctions of Isolated Populations." Paul S. Martin, Steven M. Stanley, and Thomas E. Lovejoy also contribute to this insightful work.

Osterbrook, Donald E., and Peter H. Raven, eds. *Origins and Extinctions*. New Haven, Conn.: Yale University Press, 1988.
Four outstanding evolutionary scientists contemplate the possible mass destruction of many living organisms on earth at various times in earth's history in this profound work. Alan H. Guth examines the origin and evolution of the universe. George W. Wetherill describes the formation of our solar system and earth more than four-and-one-half billion years ago. Lynn Margulis examines major events in the evolution of living forms on earth. David Raup explores the possibility of an asteroid impact destroying the dinosaurs approximately sixty-five million years ago.

Whittington, H. B. *The Burgess Shale*. New Haven, Conn.: Yale University Press, 1985.
Whittington's work is an exciting account of the discovery and content of the Burgess Shale, a region in southwest Canada where geological layers in rock have preserved millions of species of invertebrate animals that lived more than five hundred million years ago. The Burgess Shale represents the single most important piece of paleontological evidence for how the first animals may have originated. The author speculates upon how these animals originated and what their fate may have been, given the extraordinary preponderance of species preserved in this unique sediment.

Chapter 10
THE FATE OF THE DINOSAURS

Allaby, Michael, and James Lovelock. *The Great Extinction: The Solution to One of the Great Mysteries of Science, the Disappearance of the Dinosaurs*. Garden City, N.Y.: Doubleday, 1983.

In this intriguing book, Allaby and Lovelock consider the possibility that Walter and Louis Alvarez may be right concerning the extinction of the dinosaurs being caused by an asteroid or cometary impact with earth sixty-five million years ago. They consider the Alvarez theory in such chapters as 2, "Clues from the Clays," 4, "Clouds, Comets, and Meteorites," and 8, "Air Pollution on the Heroic Scale." They provide a clear, objective presentation of the data supporting the theory and of the data that questions it.

Arnold, Catherine. *Dinosaur Mountain: Graveyard of the Past*. Photographs by Richard Hewett. New York: Clarion Books, 1988.

Arnold's book is a tremendous, short introduction to dinosaur paleontology for the average person. She describes ongoing research and dinosaur fossils at the Dinosaur National Monument near the Yampa River in northeastern Utah. She clearly describes the techniques of geology and paleontology. She presents a variety of dinosaur species and conclusions about dinosaurs that researchers derive from the collected fossils and imprints. Hewett's breathtaking photographs strongly support the text, particularly with respect to demonstrations of paleontological techniques.

Bakker, Robert T. *The Dinosaur Heresies: New Theories Unlocking the Mystery of the Dinosaurs and their Extinction*. New York: William Morrow, 1986.

This excellent and sometimes irreverent look at the lives and extinction of dinosaur species is written specifically for the layperson by one of the world's leading authorities on the subject, Robert T. Bakker. Bakker maintains that dinosaurs probably were warm-blooded, an idea that runs counter to most views of cold-blooded reptiles. He also presents evidence that birds are descendants of small dinosaurs and that dinosaur extinctions probably were the result of a worldwide plague. He does believe that extinctions may be periodic, although he disputes the idea of extinctions being triggered by collisions with such things as asteroids or comets.

British Museum of Natural History (R. H. Hedley, director). *Dinosaurs and Their Living Relatives*. London: Author, 1979.

This short, simple book is an outstanding introduction to dinosaur paleontology and dinosaur paleobiology for anyone with no prior exposure to paleontology and evolution. The book is mostly photographs and illustrations, although the text succinctly covers the subject matter, including an analysis of the dinosaur's demise sixty-five million years ago. Question-and-answer sections test the reader's knowledge of the dinosauria.

Colbert, Edwin H. *Men and Dinosaurs: The Search in Field and Laboratory.* New York: E. P. Dutton, 1968.
Colbert's book is a survey of dinosaur paleontological research, particularly with reference to its history and principal scientific breakthroughs. Colbert begins with Dean William Buckland and Dr. Gideon Mantell, two English scientists who helped to launch dinosaur paleontology. He describes the works of other dinosaur paleontologists, the classification scheme for dinosaurs, and major dinosaur fossil finds in North America and Asia. The book contains candid photographs of survey teams in the field and dinosaur fossils.

Czerkas, Sylvia J., and Everett C. Colson, eds. *Dinosaurs Past and Present.* Vols. 1-2. Seattle: University of Washington Press in association with the Natural History Museum of Los Angeles County, 1988.
This two-volume collection of short essays by leading dinosaur paleontologists and ecologists describes the fossil remains, various hundreds of species, metabolisms, physiologies, and fates of dinosaurs. Contributors to volume 1 include Robert T. Bakker ("The Return of the Dancing Dinosaurs"), George Collison, Martin G. Lockley, Mark Hallett, Dale A. Russell, and David D. Gillette. Contributors to volume 2 include Gregory S. Paul, John R. Horner, Kevin Padian, Stephen A. Czerkas, Phillip J. Currie, and J. Keith Rigby, Jr. Photographs, sketches, and fantastic artwork balance the short texts of each volume.

Desmond, Adrian J. *The Hot-Blooded Dinosaurs: A Revolution in Paleontology.* New York: Dial Press/James Wade, 1976.
In this interesting work, Desmond examines the 150 million years during which the dinosaurs dominated the earth in the Mesozoic Era. Chapter 1, "The Crown of Creation," describes the history of dinosaur paleontological research. Chapter 3, "The Race Is to the Swift, the Battle to the Strong," discusses the theories of dinosaur paleontologist Robert T. Bakker, who maintains that dinosaurs may have been warm-blooded based upon their gait metabolism and other evidences. Chapter 8, "The Coming of Armageddon: A Cosmic Cataclysm," examines possible causes for the dinosaurs' demise, including local galactic supernovae (exploding stars).

Halstead, L. B., and Jenny Halstead. *Dinosaurs.* Poole, Dorset, England: Blandford, 1981.
The Halsteads describe dozens of dinosaur species, the discovery of dinosaurs, and the history of dinosaur paleontology in this exciting book. Historical analyses include Oxford naturalist Robert Plot's examination of the first discovered

dinosaur bone and the works of subsequent dinosaur paleontologists Gideon Mantell and Charles Lyell. Dinosaurs are classified group-by-group in this short paperback, which contains numerous color diagrammatic representations of dinosaurs.

Hecht, Max K., J. H. Ostrom, Gunter Viohl, and P. Wellenhofer, eds. *The Beginnings of Birds*. Eichstatt, Germany: Bronner & Daentler, KG, 1985.
This state-of-the-art discussion of paleontological research contains the proceedings of the International Archaeopteryx Conference held in Eichstatt, Germany, in 1984. The included research papers examine the presumed evolution of modern birds from dinosaur ancestors such as the flying reptile *Archaeopteryx*. Alan Fedducia speculates upon "Why the Dinosaur Lacked Feathers." Walter J. Bock examines "The Arboreal Theory for the Origin of Birds." Dietrich Schaller discusses bird "Wing Evolution." Other contributors include D. Stephan Peters, Derik W. Yalden, Philip J. Regal, and Siegfried Rietschel. Outstanding fossil photographs and data are included in this tremendous volume.

Kielan-Jaworowska, Zofia. *Hunting for Dinosaurs*. Cambridge, Mass.: MIT Press, 1969.
Kielan-Jaworowska's work is both a discussion of several Polish scientific expeditions to uncover dinosaur remains and a simple introduction to dinosaurs for the layperson. Chapter 2, "Dinosaurs and Mammals," looks at characteristics of individual species of each class of vertebrates. Chapter 4, "The First Reconnaissance Expedition and Subsequent Preparations," recounts the 1963 Polish-Mongolian expedition to the Gobi Desert. Beautiful photographs accompany these portions of the text. Chapter 12, "The Cretaceous Mammals of Bayn Dzak," discusses the mammals that eventually would displace the dinosaurs following the latter's mysterious extinction sixty-five million years ago.

Kurten, Björn. *The Age of the Dinosaurs*. World University Library. New York: McGraw-Hill, 1968.
Kurten's work is a good introduction to dinosaurs for the layperson. He includes excellent photographs, illustrations, data tables, and maps to support a lively text. Chapter 2, "Earth History Before the Dinosaurs," discusses the development of vertebrates leading up to reptiles and the giant reptiles, the dinosaurs. Chapter 3, "The Dinosaurs," describes many of the principal species, their fossils, their probable biologies, and their lives. Chapter 8, "After the Dinosaurs," examines possible events contributing to their extinction and the explosive evolutionary radiation of new mammalian species approximately sixty-five million years ago.

Ostrom, John H., and John S. McIntosh. *Marsh's Dinosaurs: The Collections from Como Bluff*. New Haven, Conn.: Yale University Press, 1966.
One of the early famous dinosaur fossil excavations was made at Como Bluff in

southeastern Wyoming by Yale University paleontologist Othneil Charles Marsh. This volume recounts Marsh's discoveries in celebration of the one hundredth anniversary of Yale University's Peabody Museum of Natural History, where many of the fossil remains are now displayed or stored. This large book contains extensive summaries of research at Como Bluff, letters by scientists, and numerous sketches of large and small Jurassic dinosaurs, reptiles, and mammals. Speculations upon the lives, physiologies, and fates of the dinosaurs are included.

Paul, Gregory S. *Predatory Dinosaurs of the World*. A New York Academy of Sciences Book. New York: Simon & Schuster, 1988.
In this entertaining book, Paul describes predatory carnivorous dinosaurs, their biology, their life histories, and their possible evolutionary fate. Chapter 3 addresses "A History of Predatory Dinosaur Success and Failure, and of Their Avian Descendants," the evolution of dinosaurs into birds. Chapters 6, "Predatory Dinosaur Speed," and 7, "Warm or Cool Predatory Dinosaurs?" discuss dinosaur paleontologist Robert T. Bakker's proposal that dinosaurs may have been warm-blooded, unlike most reptiles. Chapter 8 provides an informative "Geneology and Nomenclature of Predatory Dinosaurs and Birds."

Ratkevich, Ronald Paul. *Dinosaurs of the Southwest*. Illustrations by John C. McLoughlin. Albuquerque: University of New Mexico Press, 1976.
In this excellent introduction to dinosaur species, Ratkevich examines hundreds of dinosaur species group-by-group. In chapter 1, "Before the Dinosaurs," he describes evolutionary events preceding the rise of these reptilian giants. He then examines their development during the three periods of the Mesozoic Era. In chapter 5, "The End of an Era," he discusses dinosaur extinctions and the rise of mammals. He relates dinosaur species to present species such as birds, which very probably evolved from small, flying dinosaurs. Many photographs, fossils, illustrations, and sketches accompany the text.

Regenstein, Lewis. *The Politics of Extinction*. Foreword by Cleveland Amory. New York: Macmillan, 1975.
Regenstein's book is an important call to action to save the wildlife of the earth. According to Regenstein, more than two hundred species of animals and plants have become extinct since A.D. 1600. The numbers of extinctions are accelerating as a result of human technological negligence. Humans are initiating a massive extinction of the planet's species. He highlights his arguments in such chapters as 1, "Hunting: Killing for Fun," and 2, "The Fight to Save the Ocean Mammals." He also cites endangered species such as the American wolf and cougar and the Australian kangaroo.

Seeley, H. G. *Dragons of the Air: An Account of Extinct Flying Reptiles*. London: Methuen, 1901.

Seeley, a student of the nineteenth-century paleontologist Adam Sedgewick, provides a dated, but extremely accurate and informative, discussion of paleontology, especially with reference to the probable bird ancestors, the flying dinosaurs called pterodactyls (such as *Archaeopteryx*). In chapter 3, "A Reptile Is Known by Its Bones," Seeley provides an enthralling discussion of methods in paleontology. In chapter 5, he recounts the "Discovery of the Pterodactyle." He examines "The Plan of the Skeleton" in chapter 8. He speculates upon flying dinosaur origins and behavior in chapters 12, "Evidence of the Animal's Habits from Its Remains," and 18, "How Pterodactyles May Have Originated."

Sibley, Charles G., and Jon E. Ahlquist. *Phylogeny and Classification of Birds: A Study in Molecular Evolution*. New Haven, Conn.: Yale University Press, 1991.
Sibley and Ahlquist's book is a collection of molecular evolutionary data that impressively relates approximately 1,700 different bird species. DNA, molecular biology, and the biochemical techniques (such as DNA nucleotide sequencing, restriction endonuclease cleavage of DNA, and protein amino acid sequencing) used to study the possible evolutionary relatedness between the various bird species are described. Fossil and morphologically derived phylogenies of bird species are provided for comparison to molecular phylogenies.

Thomas, Roger D. K., and Everett C. Olson, eds. *A Cold Look at the Warm-Blooded Dinosaurs*. American Association for the Advancement of Science Selected Symposium 28. Boulder, Colo.: Westview Press, 1980.
Robert T. Bakker's hypothesis that the dinosaurs were warm-blooded, not cold-blooded like other reptilian species, is the focus of this collection of scientific papers, which debate this issue. John H. Ostrom presents "The Evidence for Endothermy" in chapter 1. Neil Greenberg describes "Physiological and Behavioral Thermoregulation in Living Reptiles" in chapter 5. Nicholas Hotton III presents "An Alternative to Dinosaur Endothermy: The Happy Wanderers" in chapter 11, the title obviously referring to the dinosaur gait and metabolism, which are the primary evidences supporting dinosaur warm-bloodedness. Robert T. Bakker caps the volume with a 110-page-long chapter 12, "Dinosaur Heresy—Dinosaur Renaissance: Why We Need Endothermic Archosaurs for a Comprehensive Theory of Bioenergetic Evolution."

Weishampel, David B., Peter Dodson, and Halszka Osmolska, eds. *The Dinosauria*. Berkeley: University of California Press, 1990.
This seven-hundred-page, detailed work contains virtually everything that is known about the hundreds of diverse species of dinosaurs. In chapter 2, "Dinosaur Paleobiology," Peter Dodson, Walter P. Coombs, Jr., James O. Farlow, and Leonid P. Tatarinov discuss our knowledge of dinosaur physiology and ecology; they also consider an extraterrestrial explanation for dinosaur extinctions. In chapter 3, "Dinosaurian Distribution," David B. Weishampel examines the

prevalence of dinosaurs on earth from 225 million years ago until 65 million years ago. Dodson, Weishampel, J. S. McIntosh, P. M. Galton, and many other scientists describe the various classes of dinosaur in twenty-six chapters devoted to dinosaur taxonomy. Many diagrams, sketches, and charts accompany these chapters.

Weyler, Rex. *Song of the Whale*. Garden City, N.Y.: Anchor Press/Doubleday, 1986.
Weyler's book is written to encourage greater conservation efforts to protect intelligent whale species from extinction in the face of the laws of several governments and the whaling industry. He cites the important work of the international organization Greenpeace and the questionable politics of the United States, the Soviet Union, and Japan concerning whaling restrictions. He enhances his case with descriptions of several individual whales, including Skana and Nicola.

Wilford, John Noble. *The Riddle of the Dinosaurs*. Drawings by Douglas Henderson. New York: Alfred A. Knopf, 1986.
In this exciting book, Wilford provides the average reader with an excellent account of dinosaur paleontology, major theories, and the scientists behind the discoveries. In chapter 5, "Early Bird," he examines the flying dinosaur *Archaeopteryx*, a possible ancestor of modern birds. Several chapters are devoted to major fossilized dinosaur finds in North America, South America, Europe, Asia, and Australia. He discusses the work and theories of controversial dinosaur paleontologist Robert T. Bakker in chapter 12, "Living Descendants?" Chapters 15, "Out with a Bang?" and 16, "Out with a Whimper?" debate the asteroid impact theory of dinosaur extinction proposed by Walter and Louis Alvarez. Other chapters examine the social lives of dinosaurs and cycles of extinction.

Chapter 11
THE DEVELOPMENT OF INTELLIGENCE IN MAMMALS

Cetaceans

Andersen, Harold T., ed. *The Biology of Marine Mammals*. New York: Academic Press, 1969.
Andersen has compiled a series of essays dealing with the physiology and behavior of such mammals as whales, dolphins, and seals. A number of prominent marine biologists have contributed to this work, including Claude L'Enfant, Jan Jansen, Richard J. Harrison, Kenneth S. Norris, William E. Evans, and Jarvis Bastian. Included scientific papers deal with the blood physiology of cetaceans, their body temperature regulation, their cardiovascular adjustments to deep diving, their nervous systems, and their reproductive cycles. Kenneth Norris describes dolphin "Echolocation" in chapter 10. Evans and Bastian discuss cetacean "Communication" in chapter 11.

Burton, Robert. *The Life and Death of Whales*. New York: Universe Books, 1973, 1980.
Burton's book is a concise summary of the lives and behaviors of whale species. He describes their unique evolution from land animals during the past seventy million years, their interesting life cycles, and their immense intelligence. He describes their unusual sensory capabilities and communication skills in chapter 4, "Submarine Senses." Chapter 6, "Behavior: Routine and Incredible," describes the complex social interactions between members of a tightly bound species. Chapter 7, "Birth and Death," describes their lengthy life cycles. The latter part of the book is devoted to the terrible, wasteful decimation of whales by the whaling industry.

Cousteau, Jacques-Yves. *The Whale: Mighty Monarch of the Sea*. Translated by J. F. Bernard. Garden City, N.Y.: Doubleday, 1972.
This outstanding book, written by the foremost marine scientist and explorer of modern times, is a story of the many diverse species of the whale, their characteristics, life cycles, and behavior. Based upon many years of extensive study throughout the world, Cousteau presents a simple, yet vivid, picture of these intelligent marine animals. Chapter 3, "When Whales Travel," is an informative presentation of whale behavior. Chapter 5, "They Talk, They Sing—And They Listen," demonstrates their intelligence and detailed "language." Chapter 11, "A Time for Respect," emphasizes the need for humans to protect the endangered whales and to end the senseless slaughter of these beautiful animals. Fantastic photographs and illustrations support the text.

Cousteau, Jacques-Yves, and James Dugan. *The Living Sea*. New York: Harper & Row, 1963.
In this exciting book, Cousteau and Dugan present more of Cousteau's explorations aboard the scientific survey ship *Calypso*. Cousteau reiterates throughout the book the beauty and diversity of ocean life, the susceptibility of ocean life to human activities (such as overfishing or pollution), and the need for humanity's awareness and protection of sea life. He describes a variety of species, including dolphins and whales, in his account. Many beautiful photographs supplement the text.

Cousteau, Jacques-Yves, with Frederic Dumas. *The Silent World*. New York: Harper & Row, 1953.
Cousteau's captivating early work of the sea for the general public arouses one's appreciation for the beauty and fragility of the oceans that constitute more than two-thirds of the earth's surface. Based upon his extensive voyages around the world, Cousteau weaves an exciting account of undersea life, including the activities of intelligent and valuable creatures such as dolphins, whales, seals, and sea lions. He breaks down the mysteries of the deep in such chapters as 10, "Sea Companions," and 11, "Monsters We Have Met."

Fichtelius, Karl-Erik, and Sverre Sjolander. *Smarter than Man? Intelligence in Whales, Dolphins, and Humans*. New York: Pantheon Books, 1972.
Two Swedish scientists, physician Fichtelius and animal behaviorist Sjolander, collaborate in this interesting general work describing experiments in dolphin intelligence. The authors argue that dolphins and other cetaceans have a higher intellectual capacity than do humans. Evidence supporting this proposition include the larger cetacean cerebral cortex (center of intelligence in the brain), elaborate cetacean communication systems, and cetacean behavior. Fichtelius and Sjolander make a concerted plea for intelligent humans to preserve the earth's environment and its many animal and plant species, including the many threatened cetaceans (such as whales).

Herman, Louis M., ed. *Cetacean Behavior: Mechanisms and Functions*. New York: John Wiley & Sons, 1980.
Herman's outstanding book is an in-depth, but very readable, survey of dolphin and whale behavior patterns. The scientific essays included in this volume are written by numerous authorities in mammalian marine biology. William W. Dawson describes "The Cetacean Eye" in chapter 2. Kenneth S. Norris and Thomas P. Dohl describe "The Structure and Function of Cetacean Schools" in chapter 5. Herman studies the "Cognitive Characteristics of Dolphins" in chapter 8. Excellent illustrations and diagrams support the text.

Lilly, John C. *Communication Between Man and Dolphin: The Possibilities of Talking with Other Species*. New York: Crown, 1978.
In this controversial, innovative work, neurophysiologist Lilly presents scientific evidence indicating that dolphins are more intelligent than humans. He bases this argument upon the intricate communication patterns of dolphins and the larger dolphin cerebral cortical brain size. He outlines these views in chapter 1, "The Development of the New Beliefs About Dolphins," and in chapter 6, "Communication by Means of Sounds: The Twin Cases of Cetaceans and Humans." He also speculates upon "The Rights of Cetaceans Under Human Laws" in chapter 15, here referring to the abuse of cetaceans by humans. He also addresses "The Possible Existence of Non-Human Languages" in chapter 16.

Mowat, Farley. *A Whale for the Killing*. Boston: Little, Brown, 1972.
Mowat's work is a touching study of the whale that he and his colleagues named Moby Joe. They track the whale along the North Atlantic coast of Newfoundland and unsuccessfully attempt to save it when it becomes ill. Throughout this tale, Mowat emphasizes the intelligence of Moby Joe and other cetaceans and the need for humans to understand these and other intelligent animals. He cites the controversial theories of Dr. John Lilly concerning exceptional cetacean intelligence in this excellent environmental book.

Nollman, Jim. *Dolphin Dreamtime: The Art and Science of Inter-Species Communication*. New York: Bantam Books, 1987.
Nollman's popular best-seller, written specifically for a general audience, describes cetacean intelligence and stresses the importance for humans to revise the way that they think about other animal species. Nollman, an authority on species communication, discusses novel techniques for studying other species, in the process demonstrating that intelligence is an attribute possessed by many types of animals. He maintains that many species, especially the cetaceans, have high intelligence comparable to human intelligence.

O'Barry, Richard, with Keith Coulbourn. *Behind the Dolphin Smile*. Chapel Hill, N.C.: Algonquin Books, 1989.
O'Barry's work is an entertaining, semiscientific study of dolphins, their intelligent behavior, and their interactions with humans, especially with the author, with fishers, and with other mariners. He describes the training of dolphins by the United States Navy for purposes such as locating mines. He also examines the work of Dr. John Lilly, a controversial scientist who researches dolphin communication and intelligence and who insists that dolphins are far more intelligent than humans. O'Barry also takes a behind-the-scenes look at the filming of the *Creature from the Black Lagoon*.

Payne, Roger, ed. *Communication and Behavior of Whales*. AAAS Selected Symposium 76. Boulder, Colo.: Westview Press, 1983.
This lengthy assembly of scientific papers, all presented at an American Association for the Advancement of Science symposium on whale communication, takes an objective look at the complex communication systems utilized by whale species. The symposium scientists analyze the composition and patterns of humpback whale songs, songs whose patterns change every year, group-specific dialects within killer whales, and the composition of whale messages. Contributing scientists to this volume include Deborah A. Glockner, George Nichols, and Christopher W. Clark.

Purves, T. E., and G. E. Pilleri. *Echolocation in Whales and Dolphins*. New York: Academic Press, 1983.
In this insightful book, Purves and Pilleri demonstrate the elaborate high-pitch vocal communication skills of the dolphin, one of the most intelligent animals on earth. The authors illustrate the physics and the physiology of dolphin communication, as well as the applications of such communication (such as sending messages or immobilizing prey). They describe studies into dolphin "language" and skull architecture in chapter 2, "The Anatomy of Phonation in Cetaceans." They discuss "Dolphin Sounds" in chapter 4 and the special case of "Phonation in the Blind Indus Dolphin" in chapter 7. "The Ear of Cetaceans" is examined in chapter 8.

Sales, Gillean, and David Pye. *Ultrasonic Communication by Animals*. London: Chapman & Hall, 1974.
In this book, Sales and Pye describe ultrasonic (high frequency) vocal communication and signaling between individuals of various animal species. Chapters 1, "Definition of Ultrasound," and 2, "Methods of Detection," describe the physical principles behind ultrasonic transmissions. Several chapters are devoted to the uses of ultrasonic communication by insects (such as crickets) and rodents (such as bats or mice) for mating, territorial displays, navigation, and capturing prey. Chapter 8, "Vertebrate Groups: Cetaceans," discusses the uses of ultrasound in the intelligent communicative messages of dolphins and whales.

Satler, Helen Roney. *Whales: The Nomads of the Sea*. Illustrated by Jean Day Zahlinger. New York: Rothrop, Lee, and Shephard, 1987.
Based upon the author's voyages aboard the ship *Tradition*, the various types of dolphins are sequentially presented, species-by-species, in this simple, entertaining book. Satler describes the physical features, life cycles, and behavioral aspects of each dolphin species with great clarity. Zahlinger's outstanding artwork supplies illustrations of each species to accompany Satler's vivid descriptions. The book is a valuable introduction to dolphins for the novice.

Schusterman, Ronald J., Jeanette A. Thomas, and Forrest G. Wood, eds. *Dolphin Cognition and Behavior: A Competitive Approach*. Hillsdale, N.J.: Lawrence Erlbaum, 1986.
The scientific essays within this important work examine the morphology of the dolphin brain, chemical and auditory communication, signaling behavior and its relationship to the social organization of dolphins, and the evolution of dolphin behavior. In chapter 7, E. W. Menzel, Jr., asks, "How Can You Tell If an Animal Is Intelligent?" J. L. Eisenberg studies "Dolphin Behavior and Cognition: Evolutionary and Ecological Aspects" in chapter 12. Douglas G. Richards observes "Dolphin Vocal Mimicry and Vocal Object Labelling" in chapter 13.

Thomas, Warren D., and Daniel K. Kaufman. *Dolphin Conferences, Elephant Midwives, and Other Astonishing Facts About Animals*. Los Angeles: Jeremy P. Tarcher, 1990.
In this entertaining animal behavior book for the layperson, Thomas and Kaufman present brief sketches (many involving mammals) of unusual animal behavior, much of which is intelligent. They discuss the scientific and evolutionary basis for many of these phenomena, which include body language, elaborate cetacean vocalizations, curious facts about many species, and dominance hierarchies (pecking orders) among individuals of various animal species.

Primates

Bourne, Geoffrey H. *The Ape People*. New York: G. P. Putnam's Sons, 1971.
Bourne's entertaining study of chimpanzees and their behavior is a very personal look at these intelligent animals. Bourne treats chimpanzees as people, people of another species who should receive our respect. He outlines these views in chapter 1, "Bobby," and chapter 2, "Ape People as Pets." He also looks at chimpanzee social behavior, particularly with reference to the work at the Yerkes Primate Research Center in Georgia. Chapter 13, "Ape People and Human People: Are They Related?" examines astronaut chimpanzees, comparisons between human infants and chimpanzees in learning behavior, and the brilliant experiments of R. Allen Gardner, Beatrice Gardner, and Roger Fouts in teaching American Sign Language to the chimpanzee Washoe.

Bright, Michael. *Animal Language*. Introduction by David Attenborough. Ithaca, N.Y.: Cornell University Press, 1985.
Bright's scientific, but very readable, study of animal communication traces the development of "language" through several vertebrate animal species, most notably bird and mammalian species, two groups that have high social bonding. Chapters 5, "Bird Calls and Songs," and 8, "Danger," describe the uses of avian communication. Chapter 2, "Singing Whales," describes the elaborate communi-

cation patterns of cetaceans. Chapter 14, "Primate Territories," discusses ape social and communicative behaviors.

Crail, Ted. *Apetalk and Whalespeak: The Quest for Interspecies Communication.* Los Angeles: Jeremy P. Tarcher, 1981.
In this thought-provoking book, Crail examines mammalian intelligence and some controversial theories surrounding such intelligence. Included in this discussion is chapter 3, "Dr. Lilly and the Dark Gulf of the Unknown," in which neurophysiologist John Lilly ranks intelligence at nine levels ranging from simple cell (level one) to "iso-humans" (that is, us, at level eight) to cetaceans (level nine), thereby making dolphins and whales more intelligent than humans. Chapter 7, "Chimpanzees: Experiences in Three Languages," discusses the brilliant work of the Gardners, David Premack, Roger Fouts, and the Hayeses in demonstrating chimpanzee intelligence and communicative capacity.

Fossey, Dian. *Gorillas in the Mist.* Boston: Houghton Mifflin, 1983.
This outstanding work, written by the late gorilla research scientist, spawned a major motion picture. In this book, Fossey outlines the scientific results of her lengthy observations of the mountain gorillas of Rwanda in eastern Africa. Chapter-by-chapter discussions of major events in various gorilla families are chronicled, including their social structure, behavior, and communication. Chapter 11, "Decimation by Poachers: Group 4," describes the senseless destruction of these intelligent creatures by human greed.

Goodall, Jane. *The Chimpanzees of Gombe: Patterns of Behavior.* Cambridge, Mass.: Harvard University Press, 1986.
Goodall's work is a thorough, but clear, analysis of the behavior of one group of chimpanzees observed from 1960 to 1986 at the Gombe Research Centre in Tanzania. Chapter 2, "The Mind of the Chimpanzee," deals with chimpanzee learning, insight, and problem solving. Chapter 6, "Communication," looks at visual, vocal, tactile, and olfactory means of contact between chimpanzees. Chapter 8, "Relationships," observes individual interactions between chimpanzees such as Figan, Goliath, Humphrey, Flo, Fifi, Flint, and Passion. Several astounding discoveries are cited, including the use of tools and the capacity to eat meat and engage in warfare and even cannibalism (as evidenced by the female Passion). The accompanying photographs are superb.

___________. *In the Shadow of Man.* Photographs by Hugo van Lawick. 1971. Rev. ed. Boston: Houghton Mifflin, 1988.
This more in-depth view of Goodall's research with the chimpanzees of Gombe is a very effective demonstration of the intelligence and personalities of chimpanzees, the fact that they are people of another species. Chapter 2, "Early Days," recounts the beginning of her research during the early 1960's with the help of

Louis Leakey. Chapter 9, "Flo and Her Family," describes the strong social bonding between the members of the family mothered by the chimpanzee Flo. Chapter 20, "Man's Inhumanity," discusses the genetic similarities between humans and chimpanzees and the threat of human technology encroaching upon the last wildlife refuges in the world. Outstanding photographs accompany the text.

___________. *My Friends: The Wild Chimpanzees*. Photographs by Hugo van Lawick. Washington, D.C.: National Geographic Society, 1967.
This book is an early look at Goodall's triumphant study of chimpanzee behavior in the wild at the Gombe Research Centre in Tanzania. With her unique, compassionate style of writing, Goodall describes the behaviors of these intelligent human relatives, including their acceptance of her presence in their midst, the lives of individual chimpanzees such as David Greybeard, and the family of Flo, Fifi, and Flint. Chapter 7, "Only the Beginning," is very prophetic for her work, which began in 1960 and continues to today. Beautiful photographs accompany the text.

___________. *Through a Window: My Thirty Years with the Chimpanzees of Gombe*. Boston: Houghton Mifflin, 1990.
This wonderful book is Goodall's autobiography of her first thirty years of research with the chimpanzees of Gombe on the shore of Lake Tanganyika. She devotes entire chapters to single individuals and the events of their lives, to chimpanzees such as Goliath, Mike, Figan, Fabin, Goblin, Flo, Fifi, Flint, David Greybeard, and Jomeo. In many ways, this book is a tribute to these individuals and represents the most profound interaction between two different species on record. She summarizes the major results of her extensive studies, and she concludes the book with a plea for the ethical treatment of animals throughout the world, especially toward "pets" and research animals.

Griffin, Donald R. *Animal Thinking*. Cambridge, Mass.: Harvard University Press, 1984.
In this important book, Griffin argues that current scientists are biased against animal thinking. He utilizes evidence from ethology, psychology, and neurobiology to argue that animals do have consciousness and that they think. He defines consciousness in chapter 1, "What Do Animals Think?" He describes mammalian and avian "Tools and Engineering" in chapter 6, "Scientific Evidence of Animal Consciousness" in chapter 7, and "Symbolic Communication" in many animal species in chapter 9.

___________. *The Question of Animal Awareness: Evolutionary Continuity of Mental Experience*. New York: The Rockefeller University Press, 1976.
In this extraordinary book, Griffin examines communicative mechanisms between

individuals of a variety of animal species. He also challenges our preconceptions of what constitutes language, intelligence, and consciousness. Chapter 2, "The Versatility of Animal Communication: Sign Language and Apes, Symbolic Communication Between Insects, Skepticism About Symbolic Communication by Honeybees," examines famous studies in animal communication research. Chapter 3, "Is Man Language?" and chapter 4, "Are Animals Aware of What They Are Doing?" address important problems in animal neurophysiology and the nature of consciousness. Chapter 6, "The Adaptive Value of Conscious Awareness," looks at the evolutionary advantages of self-awareness in animal species.

Jay, Phyllis C., ed. *Primates: Studies in Adaptation and Variability*. New York: Holt, Rinehart and Winston, 1968.
Jay's book is a compilation of scientific papers on primate behavior and communication. In essay one, K. R. L. Hall describes the "Social Organization of the Old World Monkeys and Apes." Hall also recounts "Tool Using Performances as Indicators of Behavioral Adaptability" and "Aggression in Monkey and Ape Societies" in essays four and five, respectively. John Frisch describes "Individual Behavior and Intertroop Variability in Japanese Macacques," essay nine. Jane B. Lancaster discusses "Primate Communication Systems and the Emergence of Human Language" in essay sixteen.

Jolly, Allison. *The Evolution of Primate Behavior*. New York: Macmillan, 1972.
Jolly's work is a detailed, but easy to follow, presentation of all known aspects of primate behavior from the Old and New World monkeys to the great apes. Chapter 1, "Phylogeny and Ecology," includes a taxonomical arrangement of primate relatedness. Chapter 2, "Habitat and Locomotion," discusses primate movements and their worldwide distributions (with an accompanying map). Chapter 17, "Cognition," recognizes consciousness and intelligence in chimpanzees, especially with respect to the American Sign Language experiments with the chimpanzee Washoe.

Kevles, Bettyann. *Thinking Gorillas: Testing and Teaching the Greatest Ape*. New York: E. P. Dutton, 1980.
This interesting work is a presentation of several famous experiments involving human intercommunication with gorillas. The book begins with Richard L. Gardner's 1890's study of the gorilla Othello in the New York City Central Park Zoo, research that eventually drove him to the African Congo to study gorilla language in the wild. The works of Arthur Riopelle with Snowflake and of Robert Yerkes with Miss Congo similarly are described. Recent elaborate studies of gorillas in the wild by Dian Fossey are chronicled. Furthermore, the work of Dr. Francine (Penny) Patterson in teaching American Sign Language to the gorilla Koko is highlighted.

Kohler, Wolfgang. *The Mentality of Apes*. Translated by Ella Winter. New York: Liveright, 1925, 1948, 1973.

In this early look at the study of chimpanzee and gorilla intelligence, Kohler demonstrates their great alacrity and mentality through various experiments performed during the 1920's at the Prussian Academy of Sciences Research Station. Insight and intelligence experiments are described, along with the successful manipulation of objects, experiments that many primate researchers still use today. Kohler also examines the psychology of chimpanzees.

Linden, Eugene. *Apes, Men, and Language*. New York: Saturday Review Press/E. P. Dutton, 1974.

In this fascinating book, Linden presents the brilliant experiments of R. Allen Gardner, Beatrice Gardner, and Roger Fouts in teaching American Sign Language (Ameslan) to the chimpanzee Washoe and to other chimpanzees. Chapters 1, "The Problem: A Chimpanzee That Swears," and 3, "A Tentative Comparison of Child and Chimpanzee," critically examine whether or not the chimpanzee can think analogously with humans. Chapter 9, "One Man's Description of Language," discusses scientist Charles Hockett's seven criteria for language and his insistence that apes cannot utilize language.

Mery, Fernard. *Animal Languages*. Translated by Michael Ross. Westmeade, Farnborough, Hampshire, England: Saxon House, 1975.

Mery's book is an outstanding introduction to animal communication for the layperson, with special emphasis upon chimpanzee communication. Early chapters deal with such topics as insect communication schemes (chapter 2) and amphibian mating calls and other forms of communication (chapter 5). Chapter 7, "The Six Thousand Languages of Mammals," illustrates the extensive patterns of communication developed by these highly social species. Chapter 12, "Almost Human: Communication Between Monkeys and Men," describes the work of R. Allen and Beatrice Gardner in teaching American Sign Language to the chimpanzee Washoe and attempts at communicating with dolphins by Dr. John C. Lilly.

Peterson, Dale. *The Deluge and the Ark: A Journey into Primate Worlds*. Foreword by Jane Goodall. Boston: Houghton Mifflin, 1989.

Peterson's work is an unusual personal and ecological voyage into South America, where he examines the encroachment of human civilization into wilderness areas, the destruction of rain forests, and the decimation of more than a dozen intelligent monkey species in these areas. He considers "Hunger" (chapter 4), "Food" (chapter 5), "Children" (chapter 9), and "Wood" (chapter 10) to be among several driving factors that are causing Third World countries to destroy wilderness areas and the species within these areas.

Poole, Trevor B. *Social Behavior in Mammals*. New York: Chapman & Hall, 1985.
In this introduction to the intricacies of mammalian social organization, Poole looks at the physiological and behavioral aspects of sociality in a variety of mammalian species, all of which are known for strong social bonding of individual members. Chapter 2, "Senses and Social Communication," looks at mechanisms of acoustic and visual signaling among mammals. Chapter 6, "An Order-by-Order Synopsis of Social Behavior," includes extensive references to cetacean and primate behavioral patterns in evolution. Numerous models and diagrams support the text.

Radner, Daisie, and Michael Radner. *Animal Consciousness*. Buffalo, N.Y.: Prometheus Books, 1989.
In this provocative book, the Radners debate the question of whether or not animals can think and can have consciousness. They approach this problem from a variety of angles, including the use of Kant's two tests: the language and action tests. Chapter 3, "Senseless Brutes," and chapter 4, "God and the Beast Machine," examine scientific and public misconceptions concerning animal behavior and intelligence. Chapter 7, "Animal Communication: Teaching Language to Apes," looks at animal central nervous system physiologies and the breakthrough behavioral work of R. Allen and Beatrice Gardner in teaching American Sign Language (Ameslan) to the chimpanzee Washoe and to other chimpanzees. These studies make a compelling argument in favor of many animal species having consciousness.

Rumbaugh, Duane M., ed. *Language Learning by a Chimpanzee: The LANA Project*. New York: Academic Press, 1977.
This collection of scientific papers and essays from primate language researchers at the Yerkes Primate Research Center in Georgia is very scientific but clear. The LANA project involved the communication between chimpanzees and humans via a computer console; the chimpanzees were taught to push certain buttons on the computer to ask questions or to convey messages. In chapter 4, "The LANA Project: Origin and Tactics," Rumbaugh, Harold Warner, and Ernst von Glassersfeld discuss the nature and objectives of the project. In chapter 12, "Conversations with LANA," Timothy A. Gill describes some of the results of the study.

Schiefelbusch, Richard L., and John H. Hollis, eds. *Language Intervention from Ape to Child*. Baltimore: University Park Press, 1979.
This collection of prepublished scientific journal articles presents breakthrough experiments in the history of animal behavior and animal language research. The articles deal with the nature of language and the mechanisms of communication between individuals of various species. Numerous articles deal with intelligent communicative behavior in chimpanzees, including the breakthrough experiments of R. Allen and Beatrice Gardner, Roger Fouts, Robert M. Yerkes, and Henry

W. Nissen. Other contributors include David Premack, Joseph Guch, Charity O'Neil, Sam Glucksberg, Robert M. Kraus, and Richard Schiefelbusch.

Smith, Euclid O., ed. *Social Play in Primates*. New York: Academic Press, 1978.
This work, published in association with the Yerkes Primate Research Center, is a compilation of informative articles on the evolution of intelligence and the role of play behavior in the construction of primate social structures. Donald Symons writes on "The Question of Function: Dominance and Play." John D. Baldwin and Janice I. Baldwin discuss "Reinforcement Theories of Exploration, Play, Creativity, and Psychosocial Growth." Other articles concentrate upon play behavior within certain species of primates.

Smuts, Barbara B. *Sex and Friendship in Baboons*. New York: Aldine, 1985.
In this well-written, scientific book, Smuts examines in detail the complicated social behaviors of baboon troops, including communication, visual displays, social organization, dominance hierarchies, and mating rituals. She includes two chapters on friendship behavior in baboons, chapter 4, "Defining Friendship," and chapter 5, "What Made Friends Special." Chapter 7, "Male-Male Competition for Mates," looks at male baboon territorial behavioral strategies to obtain and dominate females. Chapter 10, "Comparative Perspectives," ties in baboon behavior with human and great ape behaviors.

Snowden, Charles T., Charles H. Brown, and Michael R. Petersen, eds. *Animal Communication*. New York: Cambridge University Press, 1982.
This work is a comprehensive collection of essays on primate communication, from monkey species to the great apes to humans, written by leading authorities in the field of primate languages. Various aspects and mechanisms of communication in primates are discussed, including olfaction (smell), facial behavior, auditory localization, visual signs, and vocalizations. The anatomy and physiology of primate vocal and auditory structures are examined. Scientists contributing to this volume include Peter M. Waser, David A. Goldfoot, and Craig Dielert.

Symons, Donald. *Play and Aggression: A Study of Rhesus Monkeys*. New York: Columbia University Press, 1978.
Symons' book is a theoretical study of the behavior of rhesus monkeys, their social structure, and their communication strategies, and it makes comparisons of rhesus monkey behavior to that of the higher primates, including humans. He cites numerous experimental studies to support his findings, in which he relates how the infant play of rhesus monkeys is a mechanism of establishing dominance arrangements within the existing social structure. He arrives at these results in such chapters as 7, "Play and the Development of Skill," and 10, "Speculations on Play, Sports, and War."

Temerlin, Maurice K. *Lucy, Growing Up Human: A Chimpanzee Daughter in a Psychotherapist's Family*. Palo Alto, Calif.: Science and Behavior Books, 1975. This book is a summary of a famous animal behavior/communication experiment conducted by Temerlin and his family with the chimpanzee Lucy. Lucy was raised as a human infant under the observation of the Temerlins. The experiment was similar to the late 1940's, early 1950's raising of the chimpanzee Vicki by Keith and Catherine Hayes, whose experiments Temerlin recounts. Chapter 7, "Can Animals Talk?" is answered in the affirmative with the incredible experiments of R. Allen and Beatrice Gardner, who taught the chimpanzee Washoe more than one hundred words of American Sign Language during the late 1960's and early 1970's.

Chapter 12
HUMAN ORIGINS AND EVOLUTION

Andrews, P. J., C. B. Stringer, and Maurice Wilson. *Human Evolution: An Illustrated Guide*. New York: Cambridge University Press, 1990.
Andrews and Stringer, two leading anthropologists, combine to produce an extremely informative description of human evolution from thirty million years ago to the present. The evolution of primates such as *Aegytotithecus*, *Dryopithecus*, and the *Australopithecines* is described en route to more recent human species such as ourselves, *Homo sapiens*. Wilson provides excellent illustrations to enhance the quality of this book.

Ardrey, Robert. *African Genesis: A Personal Investigation into the Animal Origins and Nature of Man*. New York: Atheneum, 1961.
This book, which is dedicated to South African naturalist Eugene Marois, is a summary of the last five hundred million years of animal evolution, with particular reference to the origin of humans within the last one million years. Chapter 3, "The Society of Animals," describes the behavior of animals such as red ants and tigers. The role of pecking orders in animal evolution is discussed. Ardrey presents the archaeological evidence for the evolution of humans from the works of such authors as Louis Leakey and Raymond Dart. He also argues the fallacy of religious self-consciousness.

Baitsell, George Alfred, ed. *The Evolution of Man*. New Haven, Conn.: Yale University Press, 1922.
This early twentieth-century view of human evolution assembles some very scientific papers on the subject. Of particular interest to the reader are George Howard Parker's "The Evolution of the Nervous System of Man" (chapter 3) and James Rowland Angell's "The Evolution of Intelligence" (chapter 4). The book aims to incorporate all aspects of human evolutionary change, including physiological data. Information on early human species such as *Homo erectus* and *Homo sapiens neanderthalis* is provided in these informative articles. Despite the date of publication, the articles are accurate and contain good sketches and diagrams.

Bernstein, Irwin S., and Euclid O. Smith, eds. *Primate Ecology and Human Origins: Ecological Influences on Social Organization*. New York: Garland, 1979.
This collection of very interesting scientific papers attempts to link primate and human evolution based upon common features of social organization. In paper one, T. E. Rowell asks, "How Would We Know Social Organization Was Nonadaptive?" In paper nine, T. H. Cletenbrock and T. Harvey discuss "Whole

Brain Size, Population Density, and Phylogeny in Primates." B. G. Campbell describes "Ecological Factors of Social Organization" in paper twelve.

Birdsell, J. B. *Human Evolution: An Introduction to the New Physical Anthropology.* 1972. 2d ed. Chicago: Rand McNally, 1975.
Birdsell's lengthy work begins with a description of "How to Understand the Fossil Record of Evolution" (chapter 4). Chapter 9, "Clues for the Interpretation of Fossil Man," discusses the evolution of bipedalism and the worldwide distribution and migration of early humans. Chapter 11, "Men of the Lower and Middle Pleistocene," describes the origins of speech, human cannibalism, and the unusually modern-looking ER-1470 skull. Chapter 14, "A Variety of Mating Factors That Structure Human Populations," considers human assortative mating and inbreeding strategies.

Bodmer, W. F., and L. L. Cavalli-Sforza. *Genetics, Evolution, and Man.* San Francisco: W. H. Freeman, 1976.
This eight-hundred-page work is a typical genetics text, except that it places special emphasis upon evolutionary theory. The authors present the biochemical properties of DNA, the occurrence of mutations and their potential effects, chromosomal abnormalities, and natural selection for specific organismal traits within the environment. The genetics of entire organismal populations is described within the context of the Castle-Hardy-Weinberg equations. Numerous species examples are used, including comparisons of blood type systems, antibody/immunological experiments, and studies of polymorphic genes (genes affecting many traits). Part 4, "Evolution, Human Welfare, and Society," looks at human intelligence, medicine, and race on a worldwide basis.

Bowler, Peter J. *Theories of Evolution: A Century of Debate, 1844-1944.* Baltimore: The Johns Hopkins University Press, 1986.
In this book, Bowler displays the tremendous upheaval in human society created by Darwinian evolutionary theory. The implications of the theory are obvious: Humans are not supreme beings having domination over the entire earth; instead, humans and all life on earth are descended from a common ancestor. Fossil evidences of human evolution and human relatedness to apes are discussed along with the political and social ramifications. Bowler traces the debate over human origins from Darwin through the works of modern synthesis evolutionary scientists such as Theodosius Dobzhansky and Ernst Mayr.

British Museum of Natural History. *Man's Place in Evolution.* New York: Cambridge University Press, 1991.
The scientists and staff of the prestigious British Museum of Natural History have assembled a compelling amount of fossil evidence implicating humanity's evolutionary relationship to other animal species, including the great apes and

other mammals. Additional fossil evidence of the various human species that have existed during the past six million years (such as *Homo habilis* and *Homo erectus*) also is provided. The book is a clear explanation of human evolution for both the layperson and the expert.

Calvin, William H. *The Ascent of Mind: Ice Age Climate and the Evolution of Intelligence*. New York: Bantam Books, 1991.
In this provocative work, Calvin argues that human evolution, in particular the evolution of the human brain, may be linked to dramatic climatic changes on earth such as periodic ice ages. During successive ice ages over the last few hundred thousand years, the human brain has quadrupled in size. Calvin suggests that small populations of humans survived ice ages, then radiated into previously unoccupied areas that were uncovered by glacial retreat. The adaptiveness of new genes (such as genes encoding larger skull/brain capacity) could have been favored in certain isolated human populations. While making these proposals, Calvin warns us of what might happen if another ice age occurs with an overpopulation of humans on earth.

Campbell, Bernard G. *Human Evolution: An Introduction to Man's Adaptations*. Chicago: Aldine, 1966.
In this excellent anthropological work, Campbell examines the anatomical and physiological evolution of human species and their primate ancestors. Chapter 2, "Progress in Homeostasis," describes the development of warm-bloodedness, mastication, and reproductive patterns in mammalian species. Chapter 4, "Body Structure and Posture," compares the body structures of modern humans with early human species and with animal species such as gorillas and chimpanzees. Other chapters are devoted to the origin of humans, primate feeding ecology, primate behavior, and primate social behavior. Excellent diagrams and an appendix support the text.

Carlson, Elof Axel. *Human Genetics*. Lexington, Mass.: D. C. Heath, 1984.
In this book, the distinguished SUNY-Stony Brook professor describes the molecular basis of human development, disease, and evolution. Chapter 9, "Genetic Counseling," examines human genetic disorders and their diagnosis/treatment. Chapter 21, "Cancer: The Errant Cell," describes the occurrence of cancer as a somatic cell gene mutation that causes uncontrollable cellular proliferation. Chapters 23 and 28 look at "The Molecular Basis of Heredity" and "Human Evolution," respectively, from a biochemical/genetic point of view.

Cavalli-Sforza, L. L., and W. F. Bodmer. *The Genetics of Human Populations*. San Francisco: W. H. Freeman, 1971.
This nine-hundred-page volume, loaded with charts, diagrams, and extensive reference lists, is a definitive source for human genetic and evolutionary informa-

tion. The authors describe Mendelian genetic principles, gene linkage, sex linkage, chromosome anomalies, mutation and mutational rates, and the adaptability of these events to environmental pressures upon the organism. Simple mathematical models of a variety of genetic/evolutionary phenomena are presented. Examples of human genetic and evolutionary case studies include the ABO blood system, histocompatibility (tissue compatibility), cell-surface proteins, twin studies, sex ratios in populations of humans, inbreeding, progeny size distributions, and race.

Chiarelli, A. B., ed. *Comparative Genetics in Monkeys, Apes, and Man*. New York: Academic Press, 1971.
This volume is a collection of scientific papers presented by several authorities in primate evolution at a July, 1970, symposium on comparative genetics in primates and human heredity at Ernice, Sicily. The contributed articles contain useful data tables, charts, and reference studies that provide insights into human origins and human relatedness to other primate species. A. Caroline Berry and R. J. Berry examine polymorphisms in primate skeletons. Morris Goodman and colleagues study "Primate Genes and Proteins." J. Mavolwala looks at primate dermatoglyphic traits, while A. O. Carbonara studies primate antibodies/immunology and editor A. B. Chiarelli closely illustrates comparative cytogenetic similarities between primates and humans.

Chiarelli, A. B., Ann L. Koen, and G. Ardito, eds. *Comparative Karyology of Primates*. New York: Mouton, 1979.
This collection of scientific papers focuses upon the similarities between humans, the great apes, monkeys, and other primates at the molecular level. The papers presented in the volume compare chromosomes between various species, in the process obtaining striking evidence supporting the evolutionary relatedness of the various species. Comparisons of DNA nucleotide and protein amino acid sequences also are provided. Data gleaned from these reports enable the construction of molecular phylogenetic trees that illustrate the probable evolutionary descent of the various primate species. Contributors to the volume include Chiarelli, Koen, Ardito, Bodil E. Lomholt, A. Folek, D. Soudek, and Catherine Turleau.

Ciochon, Russell L., and John G. Fleagle, eds. *Primate Evolution and Human Origins*. Menlo Park, Calif.: Benjamin/Cummings, 1985.
This book is a collection of outstanding essays written by leading anthropologists and primate biologists. F. A. Jenkins, Jr., examines "Tree Shrew Locomotion and the Origins of Primate Arborealism." C. O. Lovejoy speculates upon "The Origin of Man." V. Sarich proposes "A Molecular Approach to the Question of Human Origins," an article that provides compelling evidence for human-ape common ancestry. R. L. Ciochon describes "Hominoid Cladistics and the Ancestry of Modern Apes and Humans." The many articles compare anatomy, physiology, and biochemistry between species.

Dart, Raymond A., with Dennis Craig. *Adventures with the Missing Link.* New York: Harper & Brothers, 1959.
Dart, a prominent anthropologist/archaeologist, presents the history of the study of human origins. He describes the early discovery of early humans such as *Australopithecus*, *Homo erectus*, and *Homo habilis*. He speculates on the evolution and cultures of early humans in chapters 14, "How Man Got off His Knuckles," and 16, "How the Past Reveals the Future." This book is an important contribution to our understanding of human origins and evolution from one of the leading scientists who performed pioneering work in these fields.

Dennett, Daniel C. *Consciousness Explained.* Boston: Little, Brown, 1991.
Dennett, a Tufts University cognitive scientist, describes the very nature of human consciousness in this powerful book. He proposes a new theory of consciousness and intelligence called the Multiple Drafts model, in which human thought is a product of the analysis of incoming stimuli by many different regions of the brain; any particular thought or idea is a combination of interactions from many different cerebral centers. He combines impressive philosophical explanations and scientific data to support his theory.

Dobzhansky, Theodosius. *The Biology of Ultimate Concern.* New York: New American Library, 1967.
This philosophical work, written by one of the foremost evolutionary geneticists of the twentieth century, discusses the nature of humans, human evolution, and the future possibilities of human potential. Chapter 2, "On Gods of the Gaps," describes the work of various scientists from Sir Isaac Newton to Thomas Hunt Morgan in discovering the processes of nature and of life. Chapter 4, "Self-Awareness and Death-Awareness," concentrates upon the origins of human intelligence and consciousness.

Dobzhansky, Theodosius, and Ernest Boesiger. *Human Culture: A Moment in Evolution.* Edited and completed by Bruce Wallace. Illustrations by Hans Erni. New York: Columbia University Press, 1983.
In this philosophical work, two eminent evolutionary scientists discuss patterns and mechanisms of human evolution, using important scientific studies to support their arguments. Chapter 1, "Different Forms of Natural Selection," describes the evolutionary adaptiveness of sickle-cell anemia, a gene that confers malarial resistance to heterozygote sickle-cell carriers. Chapter 4, "The Unique Position of the Human Species in Evolution," describes the evolution of human culture, language, self-awareness, and death awareness. Chapters 5, "The Biological and Cultural Evolution of Man," and 7, "Human Equality and Genetic Diversity," examine the future of human society and evolution.

Donald, Merlin. *Origins of the Modern Mind: Three Stages in the Evolution of Culture and Cognition*. Cambridge, Mass.: Harvard University Press, 1991.
In this thought-provoking and controversial work, Donald presents a dramatically alternative view as to how human intelligence arose. He proposes that cognitive powers, the ability to process and store knowledge, have proceeded through three phases in the evolution of the human species during the past two million years. The first phase was the storage of knowledge via physical acts and gestures. The second phase was the development of language, storytelling, and myths. The third phase is the symbolic representation of knowledge via the written word, art, machines, and artificial intelligence on computer. He stresses that the evolution of human cognition is a dynamic process that continually is changing.

Dubos, René. *The Torch of Life: Continuity in Living Experience*. New York: Simon & Schuster, 1962.
This short, philosophical essay is a study of humankind's origins, humanity's place within the universe, and the future of human social evolution. Dubos examines "Cosmic Cycles and Biological Progress" to illustrate the evolution of life on earth as an extension of universal evolution and humanity's role within the context of life on earth. "The Adaptability of Man" describes human evolution, both morphologically and culturally. The social nature of humans is emphasized throughout the text.

Durant, John R., ed. *Human Origins*. New York: Oxford University Press, 1989.
Durant has assembled an impressive collection of informative essays on human evolution in this volume, which includes important contributions by some of the world's leading anthropologists and paleontologists. Richard Leakey and Richard Dawkins are among the authorities who discuss the development of intelligence in hominids, the sociocultural evolution of early humans, and major physical changes in the human form.

Eldredge, Niles, and Ian Tattersall. *The Myths of Human Evolution*. New York: Columbia University Press, 1982.
In this book, two leading evolutionary scientists tear down several misconceptions about what evolution is and how it occurs. They examine a wealth of data, particularly from the fossil record, to glean an understanding of how humans have evolved during the past few million years. They point out that evolution often occurs in short, rapid spurts. Consequently, human evolution probably will see very little change during the next few tens of thousands of years, barring catastrophes, of course.

Goldsmith, Timothy H. *The Biological Roots of Human Nature*. New York: Oxford University Press, 1991.
Goldsmith, a Yale University scientist, emphasizes the causal relationship between

evolution, human genes, and the development of human intelligence and culture in this provocative book. Furthermore, he argues that human intelligence and culture can and has influenced human evolution, thus completing the cycle. Sociologists, philosophers, historians, and other observers of human nature should recognize these important interrelationships between human society, culture, and evolution, particularly as humans undergo ever-increasing changes in their cultural evolution.

Grafen, Alan, ed. *Evolution and Its Influence*. Oxford, England: Clarendon Press, 1989.
Anthropologists, biologists, and philosophers speculate upon various aspects of human social and cultural evolution in this collection of stimulating essays. The selected topics in this volume include primate and human behaviors, selection pressures in human evolution, incest taboos, human sociocultural evolution, and evolution in human art forms. Contributing scientists include A. J, Cain, Bernard Williams, and Maurice Godelier.

Guyer, Michael F. *Speaking of Man: A Biologist Looks at Man*. New York: Harper & Brothers, 1942.
In this interesting study of human evolution, Guyer looks at the biological and cultural evolution of humans. Chapter 3, "Man's Place in Nature," examines human origins and development over the past one million years. Chapter 6, "The Endocrine Control of the Body," demonstrates the similarities of hormones and their influences in the bodies of humans and other mammals. Other chapters such as "Sex" (chapter 7) and "Democracy as a Biological Problem" (chapter 8) deal with contradictions between our cultural and biological evolutions.

Haeckel, Ernst. *The Evolution of Man: A Popular Exposition of the Principal Parts of Human Ontogeny and Phylogeny*. New York: D. Appleton, 1897.
This dated, yet very accurate, discussion of evolutionary theory by a leading nineteenth-century evolutionary scientist provides extensive evidence and diagrams supporting the evolution of humans from primates. Haeckel presents evidence from comparative anatomy, comparative embryology, and physiology to argue humanity's place within the evolution of vertebrate animal species. He also recounts the history of evolutionary theory and the mechanisms of evolutionary progress.

Hanson, Earl D., ed. *Recombinant DNA Research and the Human Prospect*. Washington, D.C.: American Chemical Society, 1983.
This collection of presentation papers for a sesquicentennial symposium at Wesleyan University, Middleton, Connecticut, assimilates the ideas of scientists, thinkers, and leaders on the next step in human evolution: manipulation of the genetic material in humans and in other organisms. Paper topics include the

biochemistry of DNA, experimental techniques currently used in cloning genes, ethical considerations in recombinant DNA research, the business aspects of bioengineering, and possible regulations to control cloning and other related recombinant DNA experiments. Contributors to the volume include Salvador Luria, Philip Abelson, and Robert A. Swanson.

Herrick, C. Judson. *The Evolution of Human Nature.* Austin: University of Texas Press, 1956.
This part-philosophical, part-biological examination of human evolution concentrates primarily upon the development of human behavior and the complexity of the human brain. Herrick looks at major trends in the evolution of vertebrate animal species over the past few hundred million years. The behaviors and social advancement of many groups of species are compared and traced to the evolution of more complicated, "plastic" (adaptable) nervous systems. In primates and humans, Herrick studies nervous system evolution, the expansion of the cerebral cortex (the brain), and the psychobiology of behavior. He also examines concepts of mind and of self.

Hooton, Earnest Albert. *Up from the Ape.* New York: Macmillan, 1946.
In this lengthy, comprehensive study of human evolution, the anthropologist Hooton vividly describes the relationship of humans to other primate and mammalian species using a wealth of reference data. Part 2, "The Primate Life Cycle," describes the evolution of arboreal (tree-dwelling), quadrupedal primates to terrestrial, bipedal hominids having rearranged teeth and very little body hair. Part 3, "The Individual Life Cycle," compares and contrasts the lives of humans and other primate species. Part 4, "Fossil Ancestors and Collaterals," describes the methodologies and discoveries that led to our current picture of human origins and evolutionary patterns. Part 5, "Heredity and Race," discusses the worldwide distribution of human genetic variations. This book is a tremendous reference.

Howells, William, ed. *Ideas on Human Evolution: Selected Essays, 1949-1961.* Cambridge, Mass.: Harvard University Press, 1962.
This collection of theoretical, scientific, and philosophical papers on human evolution are reprinted from scientific journals and other publications. The many fascinating essays describe the origin of *Homo sapiens*, the phylogenetic relationship of humans to other primate species, hominid evolution, the specialized evolution of certain primate traits (such as mastication), and the social ecology of primate species. Contributors to this volume include such eminent scientists as George Gaylord Simpson, Theodosius Dobzhansky, William K. Gregory, Ernst Mayr, W. E. Le Gros Clark, Joseph B. Birdsell, and G. H. R. von Koenigswald.

____________. *Mankind in the Making: The Story of Human Evolution.* Garden City, N.Y.: Doubleday, 1959.

In this historical and evolutionary account of mankind's origins, Howell's illustrates the evolution of vertebrate animals to mammals to anthropoid apes to humans. In the process, he discusses the scientists who made important breakthroughs in uncovering humanity's past. Chapter 7, "Dart, Broom, and *Australopithecus*," describes the anthropological discoveries of early human species such as *Australopithecus*, Java man, Peking man, and *Homo sapiens neanderthalis*. Chapters 14, "*Homo sapiens* Arrives," and 18, "Races of Man," describe the appearance and spread of humanity across the planet. Excellent diagrams and photographs supplement the text.

Johnson, Donald, and Maitland Edey. *Lucy: The Beginnings of Humankind*. New York: Simon & Schuster, 1981.
Johnson and Edey's work is a clearly written, beautifully illustrated and photographed summary of the major breakthroughs in twentieth-century human archaeological research. Their work begins with a discussion of Charles Darwin, the Darwinian-Wallace Theory of Evolution, and early human fossil finds. The book progresses through the discovery of ancient human species in East Africa and Asia. It culminates with the brilliant, laborious work of Louis and Mary Leakey, the discoverers of *Homo habilis*.

Klein, Richard G. *The Human Career: Human Biological and Cultural Origins*. Chicago: University of Chicago Press, 1989.
Klein describes the evolution of primates and humans from eighty million years ago to the present. He supplies the reader with accurate paleontological and archaeological data, fossil evidence, and archaeological methods. Competing theories concerning human origins and phylogenies relating humans to ape species are discussed. He presents evidence indicating that *Homo sapiens sapiens* emerged in Africa approximately two hundred thousand years ago, then spread into Europe and Asia while outcompeting older species of humans.

Kraus, Bertram S. *The Basis of Human Evolution*. New York: Harper & Row, 1964.
This excellent survey textbook on human evolution analyzes all items of evidence that relates humans to the great apes. Kraus looks at the process of evolution, including genes, mutations, chromosome rearrangements, and the effects of these factors upon the organism in relationship to selection pressures from the environment. He presents extensive fossil evidence for human evolution. He also traces the migration of early humans from their origins in eastern Africa to other parts of the world. Comparative genetics and morphologies of gorillas and humans enable the construction of possible phylogenies for human evolution from anthropoid apes.

Landau, Misia. *Narratives of Human Evolution*. New Haven, Conn.: Yale University Press, 1991.

Landau's work is an unusual and refreshing approach to the study of evolution. She analyzes the presentation of evolutionary theory by several notable scientists, including Darwin himself. She observes that what each scientist perceives to be the principal driving force in evolution determines how the evolutionary data subsequently is interpreted. Consequently, evolutionary views can be biased and require cautious scrutiny.

Lanpo, Jia, and Huang Weiwen. *The Story of Peking Man*. New York: Oxford University Press, 1990.
Two prominent Chinese paleoanthropologists, Lanpo and Weiwen, recount the major scientific discoveries surrounding the early human that anthropologists call Peking man, who lived in what is now mainland China during the Pliocene and Pleistocene epochs. Lanpo is an authority on Peking man; he has researched Peking man fossils for more than fifty years. He and Weiwen describe fossils isolated from the Zhoukoudian region south of Beijing, discuss the history of Peking man research, and speculate upon the sociology and culture of this early human. This book is an important contribution to the human archaeological literature.

Lasker, Gabriel W. *The Evolution of Man*. New York: Holt, Rinehart and Winston, 1961.
Lasker's work is an analysis of animal evolution through the evolution of humans, with primary reference to the latter. He begins with a study of evolutionary theory and mechanisms, the DNA molecule and genes, the occurrence and effects of gene mutations, chromosome abnormalities, and the patterns of natural selection upon individual traits as well as upon entire animal populations. He surveys the animal kingdom up to the primates, whose anatomy, physiology, and behavior are carefully scrutinized. He proceeds to hominid evolution, where he presents detailed coverage of the fossil remains of early human species: Java man, Pekin man, *Homo erectus*, *Homo habilis*, and *Homo sapiens neanderthalis*. He also discusses blood group evidence for human evolution and the races of humanity. The book is well organized with good photographs.

____________, ed. *The Progress of Ongoing Human Evolution*. Detroit: Wayne State University Press, 1960.
Several leading human evolutionary scientists contribute to this volume that analyzes the physiological basis of human development during the past one million years. Paul T. Baker discusses "Climate, Culture, and Evolution," Frank B. Livingstone describes the ABO blood type system and the epidemiology of various human diseases, Frederick S. Hulse emphasizes the nervous system plasticity that has expanded and optimized the human brain, and Earle L. Reynolds examines the effects of radiation upon future human evolution. The essays provide the reader with much information regarding the genetic past and future of humanity.

Laughlin, W. S., and R. H. Osborne, introductions *Human Variation and Origins: An Introduction to Human Biology and Evolution, Readings from Scientific American*. San Francisco: W. H. Freeman, 1967.
This collection of *Scientific American* reprints consists of survey articles concerning the origin and evolution of human species. Contributing authors include Theodosius Dobzhansky, Loren Eiseley, Francis Crick, and Julian Huxley. The papers examine topics such as the genetic basis of evolution, mammalian developmental hormones, the evolution of the human prehensile (grasping) hand, the origins of speech, human genetic diseases such as sickle-cell anemia, and the archaeological studies of early human species (such as *Homo erectus*).

Le Gros Clark, W. E. *The Antecedents of Man*. New York: Harper & Row, 1959.
Le Gros Clark, a pioneer in the anthropological/archaeological studies of early humans, presents considerable evidence supporting common ancestry between the great apes and humans. Using anatomical and physiological comparisons between species, he demonstrates common descent using dentition (teeth) studies, skull analyses, limb arrangements and motions, brain size, special senses, digestive systems, and reproductive systems. Chapter 10, "The Evolutionary Radiation of the Primates," presents primate evolution of the past fifty million years.

____________. *The Fossil Evidence for Human Evolution: An Introduction to the Study of Paleoanthropology*. Chicago: University of Chicago Press, 1955, 1964.
This treatise on human evolution, written by one of the pioneers in human archaeology, examines the fossil evidence for human evolution from anthropoid apes. Le Gros Clark carefully analyzes the comparative morphologies of human, early human, gorilla, chimpanzee, and other primate skeletons. He comments on primate and hominid skull evolution. He examines a variety of fossil human skeletons ranging from *Homo erectus* to *Homo sapiens neanderthalis*. He speculates on the origin and evolutionary patterns of hominid species over the past one million years in this valuable reference.

Leakey, Louis S. B. *Adam's Ancestors: An Up-to-Date Outline of What Is Known About the Origin of Man*. London: Methuen, 1934.
This early work by the famous anthropologist/archaeologist describes early twentieth-century breakthroughs in the study of early humans. Leakey cites detailed fossil evidence for early human species. He describes their anatomical structure, their manipulation of stone tools, and the environment in which they probably lived. He makes skeletal comparisons with ape species to demonstrate the nature of human origins. He also speculates upon early human cultures.

Leakey, Louis S. B., and Vanne Morris Goodall. *Unveiling Man's Origins: Ten Decades of Thought About Human Evolution*. Cambridge, Mass.: Schenkman, 1969.

In this exciting book by two giants of twentieth-century anthropology, Leakey and Goodall present a decade-by-decade account, from 1860 to 1968, of major breakthroughs in the scientific study of human origins. The discovery of such early human fossils as *Dryopithecus*, *Australopithecus*, *Homo erectus*, *Homo habilis*, and *Homo sapiens neanderthalis* are recounted, along with false alarms such as the Piltdown controversy. This excellent contribution to the anthropological literature contains great photographs, diagrams, and a glossary.

Leakey, Louis S. B., and Jack and Stephanie Prost, eds. *Adam, or Apes: A Sourcebook of Discoveries About Early Man*. Cambridge, Mass.: Schenkman, 1971.
This book is a collection of previously published essays and scientific articles dealing with the fact of human evolution and controversies mounted by creationist opponents of evolutionary theory. Contributing authors include Leakey, the Prosts, Robert Broom, Raymond Dart, W. E. Le Gros Clark, Elwyn L. Simons, and Thomas Henry Huxley. Courtroom comments of opposing lawyers William Jennings Bryan and Clarence Darrow from the Scopes "Monkey" Trial are also included in this volume. Besides the controversy, considerable fossil evidence of early human species is presented in the scientific articles.

Leakey, Richard E., and Roger Lewin. *Origins*. New York: E. P. Dutton, 1977.
This book, written by two eminent scientists, is a modern view of how humans have evolved during the past six million years. Leakey, the son of the famous archaeologists Louis and Mary Leakey, and Lewin describe the evolution of humanlike hominids from their early beginnings to modern *Homo sapiens*. They maintain that possibly three species of humans may have existed simultaneously in eastern Africa. They hypothesize that cooperation, not necessarily competition, led to the evolution of later humans from these three early human species. The book is loaded with spectacular color photographs.

____________. *People of the Lake: Mankind and Its Beginnings*. Garden City, N.Y.: Anchor Press/Doubleday, 1978.
The prominent human archaeologists Leakey and Lewin describe human origins along the lakeshores of eastern Africa during the past few million years. Chapter 4, "A New Perspective on Human Origins," looks at fossil evidences from east Africa, the evolution of the genera *Australopithecus* and *Homo*, and the possibility that our immediate ancestors may have been killer apes. Chapter 8, "The First Affluent Society," examines early human culture and phenomena such as altruism (helping others). "The Nature of Intelligence" and "The Origins of Language" are studied in chapters 9 and 10, respectively. Excellent photographs support the text.

Lewin, Roger. *Human Evolution: An Illustrated Introduction*. New York: W. H. Freeman, 1984.

Lewin's short one-hundred-page synopsis of human evolutionary change is a strong introduction to the subject for the layperson. Despite its brevity, the book has tremendous scope. Chapter 6, "Molecules Tell Evolutionary Stories," describes the use of molecular biochemical studies to illustrate the relatedness of humans to other primate species. In chapter 7, Lewin discusses "Major Steps in Human Evolution." A succession of early human species is presented in chapters 14, "The Australopithecines," 15, "*Homo habilis*," 16, "*Homo erectus*," and 22, "The Neanderthals." Chapter 23, "Origin of Modern Humans," examines the apparent similarities of early European *Homo sapiens* to modern Europeans.

Lieberman, Philip. *Uniquely Human: The Evolution of Speech, Thought, and Selfless Behavior*. Cambridge, Mass.: Harvard University Press, 1991.
Lieberman's incredible work is an exciting analysis of human evolution and some remarkably alternative theories as to how human intelligence may have developed during the past one hundred thousand years. He argues that human language and the development of vocalization may have helped to accelerate the evolution of the human cerebral cortex. He utilizes biological, psychological, archaeological, and computer simulation evidence to support his views.

Lumsden, Charles J., and Edward O. Wilson. *Promethean Fire: Reflections on the Origin of Mind*. Cambridge, Mass.: Harvard University Press, 1983.
Two prominent evolutionary scientists combine to analyze the origins of intelligence within the human mind. Based upon the theory of sociobiology developed by Wilson, Lumsden and Wilson argue that human intelligence is part genetic, part cultural. Our species has constructed its culture to reinforce the development of mind: The ability to do that is genetic and is a unique adaptation for our species. The authors illustrate their position with many examples. Their arguments invite scientists to diversify their approach to studying human evolution.

McCrone, John. *The Ape That Spoke*. New York: William Morrow, 1991.
This discussion of human evolution concentrates on the evolution of the human brain. McCrone argues that the expansion of the human brain's abilities, not its size, led to the evolution of modern humans. He correctly points out that the approximately one hundred billion neurons constituting the human brain exhibit a considerable degree of plasticity, the ability to grow and alter growth in specified patterns. Consequently, the adaptiveness of neural plasticity in humans may have given the human brain a substantial edge over other species in the struggle for survival, thereby enhancing the success of humans within the last one hundred thousand years.

Martin, R. D. *Primate Origins and Evolution: A Phylogenetic Reconstruction*. Princeton, N.J.: Princeton University Press, 1990.
Martin's work is a lengthy, comprehensive study of primate evolution over the

past seventy million years. He describes the evolutionary changes that gave rise to the tree shrews, tarsiers, Old and New World monkeys, and the great apes. Human evolution is also described in detail. Martin cites fossil, chromosomal, and molecular evidence to support the modern classification and patterns of evolution for various primate groups. The book includes more than two hundred illustrations.

Mather, Kenneth. *Human Diversity: The Nature and Significance of Differences Among Men*. New York: Free Press, 1964.
Based upon his genetics lectures at the University College of North Wales, Mather's book is a clear presentation of human evolution in layperson's terms. He discusses various factors that have combined to influence human evolution, including the environment, genes, and mutations. He describes the effects of gene blocks called polygenes upon the evolution of certain human characteristics. He also looks at polymorphisms and sickle-cell anemia. Social evolution in primate and human societies is compared in light of the development of higher intelligence.

Maxwell, Mary. *Human Evolution: A Philosophical Anthropology*. New York: Columbia University Press, 1984.
Maxwell's book is a simple, nontraditional approach to the origins and evolution of human species. She looks at the origin of life on earth, the evolution of species leading up to modern humans, and she specifically concentrates on the evolution of particular factors that make humans what they are: intelligence, consciousness, language, and culture. The evolution of the chemistry of behavior is also described.

Mellars, Paul, and Christopher Stringer, eds. *The Human Revolution: Behavioural and Biological Perspectives on the Origins of Modern Humans*. Princeton, N.J.: Princeton University Press, 1989.
This volume is a collection of scientific papers by fifty-five leading anthropologists and human geneticists who attended a 1987 human evolution conference at Cambridge University. The papers employ the use of molecular data (DNA and protein) from human populations to substantiate the evolutionary finds of anthropologists and archaeologists. The origins of human language, analysis of human mitochondrial DNA, and possible human phyologenies are described.

Montagu, Ashley. *Human Heredity*. New York: World, 1959.
This introduction to human genetics, written by a leading evolutionary scientist, describes the physical/chemical basis of heredity within humans. Chapter 2, "In the Beginning," discusses the origin and evolution of the DNA molecule. Chapter 5, "Heredity and the Environment," describes the interplay between an individual's genes and environmental factors (such as temperature or chemicals) in the

determination of the person's physical makeup. Other chapters are devoted to human sexuality, heredity, and the effects of atomic radiation upon humans and upon subsequent human evolution.

Moore, Ruth. *Man, Time, and Fossils: The Story of Evolution.* Drawings by Sue Richert. New York: Alfred A. Knopf, 1953, 1961.
Moore's book is a historical survey of human archaeological and paleoanthropological research during the nineteenth and twentieth centuries. Part 1, "Man's Origins," includes chapters on Charles Darwin (chapters 2 to 4), Jean Baptiste de Lamarck, Hugo de Vries, Gregor Mendel, and J. B. S. Haldane and on their scientific contributions to evolutionary theory. Part 2, "Man's Buried Record," describes the archaeological discoveries of Gustav Heinrich Ralph von Koenigswald, Rene Dubois, and Raymond Dart. Part 3, "A Changed Theory of Man's Evolution," includes discussions of radioactive carbon-14 dating and the works of Louis and Mary Leakey, Willard F. Libby, Adolph Knopf, and others.

Morris, Desmond. *The Naked Ape: A Zoologist's Study of the Human Animal.* New York: McGraw-Hill, 1967.
Morris' popular and controversial book on human evolution is a major work in the field yet is very accessible to the layperson. Morris calls attention to many unusual facets of human behavior, anatomy, and physiology. For example, *Homo sapiens sapiens* is the only one of 193 primate species that does not have extensive body hair. Furthermore, Morris speculates upon various human behaviors and psychologies in relation to sex, child rearing, and aggression. This book is must reading.

Oxnard, Charles E. *The Order of Man: A Biomathematical Anatomy of the Primates.* New Haven, Conn.: Yale University Press, 1984.
Oxnard's detailed, but very clear, book relies upon comparative anatomy, comparative physiology, and the fossil record to construct a possible phylogeny of the primates. He uses mathematical models to theorize the degree of relatedness between humans and the great apes and to construct a possible scenario of human evolution from the great apes. He plays close attention to similarities in human and ape limb construction and behavior.

Pfeiffer, John E. *The Emergence of Man.* New York: Harper & Row, 1969.
This lengthy, well-researched volume observes primate and human evolution over the past thirty million years, according to fossil and molecular evidences. Chapter 1, "Primate Origins," discusses the appearance of color vision and an opposable thumb in higher monkeys and apes. Chapter 2, "The First Members of the Family of Man," describes the prehuman *Ramapithecus*, the evolution of bipedalism (walking on two feet), and tool use among chimpanzees and early humans. Chapter 10 speculates upon "The Disappearance of Neanderthal Man, the

Appearance of Modern Man." Chapter 18, "The Human Infant: A Study in Living Prehistory," provides an unusual look at comparisons of human and primate development/evolution.

Rifkin, Jeremy, in collaboration with Nicanor Perlas. *Algeny*. New York: Viking, 1983.
In this ethical treatment of a very controversial topic, cloning, Rifkin looks at the pros and cons of recombinant DNA research, a relatively new branch of biology that offers tremendous medical promises but also disastrous consequences if it is not conducted in a responsible manner. He invokes Darwinian evolution and Nietzche's philosophy to test the effects, good or bad, of our escalating technology. Part 2, "Deciding What's Natural: The Ultimate Intellectual Deception," is a stirring discussion of evolutionary ethics, biological clocks, and the impact of biotechnology. The author stresses that modern science is moving too rapidly for society to cope.

Sagan, Carl. *The Dragons of Eden: Speculations on the Evolution of Human Intelligence*. New York: Ballantine, 1977.
Sagan, in his inimitable style, presents another popular account of the methods of science in this best-selling work. He describes the biological evolution of primates and humans, the physiological development of the human brain, and the genetic changes that have contributed to the enhancement of intelligence and cognition in the human species. He places special emphasis upon the cultural evolution of humanity. He describes numerous examples of social and cultural changes in human evolution during the past one million years, some of which still exist in assorted human populations across the globe.

Spuhler, J. N., arranger. *The Evolution of Man's Capacity for Culture*. Detroit: Wayne State University Press, 1959.
Spuhler, a University of Michigan anthropologist, has brought together leading anthropologists to trace the evolution of primate and human cultures. Ralph W. Gerard discusses "Brain and Behavior," S. L. Washburn provides "Speculation on the Interrelations of the History of Tools and Biological Evolution," and Harry F. Harlow analyzes "Basic Social Capacity of Primates." Other contributors include Charles F. Hackett, Marshall D. Sahlins, and Leslie A. White.

Szalay, Frederick S., and Eric Delson. *Evolutionary History of the Primates*. New York: Academic Press, 1979.
In this important work, Szalay and Delson present the evolutionary history of the primates over the past sixty million years. They present each primate species, and list its characteristics, anatomy, physiology, and behavior. The species are arranged by family taxonomic categories. Many illustrations are provided, and

numerous photographs of teeth, skulls, and entire skeletons are provided to support common evolutionary descent between these various species.

Tattersall, Ian, Eric Delson, and John A. Van Couvering. *Encyclopedia of Human Evolution and Prehistory*. New York: Garland, 1988.
Tattersall, Delson, and Van Couvering of the American Museum of Natural History have assembled one of the most complete collections of information relating to human evolution ever assembled. More than one thousand entries describing every aspect of evolutionary theory, paleontology, archaeology, anthropology, and human evolution are included in this impressive work written by several dozen authorities on the subject matter. It is an invaluable reference for both layperson and scientist alike and has received numerous awards for excellence.

Vogel, F., and G. Rohrborn, eds. *Chemical Mutagenesis in Mammals and Man*. New York: Springer-Verlag, 1970.
This compilation of research papers, presented at a 1969 symposium in Mainz, Germany, is devoted to the analysis of genetic changes in human and mammalian populations as a result of exposure to radiation and chemical mutagens. Contributors include I. D. Adler, C. G. Arnold, U. H. Ehling, E. Gebhart, J. Krüger, P. K. Dalton, and H. Malling. The articles contain supporting graphs, charts, and some mathematical models to illustrate the potential impact of radiation and chemical mutagens upon human birth defects and subsequent human evolution.

Von Koenigswald, Gustav Heinrich Ralph. *The Evolution of Man*. Ann Arbor: University of Michigan Press, 1962.
This short but interesting history of evolution begins 2.64 billion, years ago with early bacteria. The author then proceeds through a rapid summary of major evolutionary events leading up to vertebrate animals. With vertebrates, von Koenigswald uses the fossil record and comparative anatomy to trace vertebrate evolution from fish to amphibians to reptiles to birds and mammals. With mammals, he describes the evolution of intelligence in primates and early humans to present-day *Homo sapiens*. The book contains many photographs, sketches, and data tables.

Vonnegut, Kurt. *Galapagos*. New York: Dell Books, 1985.
In this brilliantly written satirical novel, the eminent writer Vonnegut analyzes the human situation with his usual cast of pitiful, average people. His story revolves around a nature cruise to commemorate Darwin's voyages and the extinction of modern *Homo sapiens*. In the process, he accurately describes the mechanisms of evolutionary change, especially the element of chance, with dark humor. He emphasizes the irrationality of human thought, the needless destruction brought about by human greed, and the need for humanity to unite with drastically

altered views of how life should be conducted. Based upon current human behavior, Vonnegut's book argues that humans might be better off as seals (into which the descendants of his characters evolve one million years hence).

Washburn, Sherwood L., ed. *Classification and Human Evolution*. Chicago: Aldine, 1963.

This collection of essays by leading biologists, molecular geneticists, psychologists, and philosophers is an excellent survey of human evolution. The informative articles survey possible human phylogenetic relationships to other primate species such as gorillas, orangutans, and chimpanzees, the behavior and psychology of primate species, sexual differences in primates, hominid locomotion, and molecular anthropology. Contributors to this volume include such eminent scientists as George Gaylord Simpson, John Napier, Adolph Schultz, Morris Goodman, Ernst Mayr, and Theodosius Dobzhansky. Each article contains important human evolutionary data and diagrams.

Washburn, Sherwood L., and Phyllis C. Joy, eds. *Perspectives on Human Evolution I*. New York: Holt, Rinehart and Winston, 1968.

Another collection of scientific papers by eminent evolutionary scientists from several different but related disciplines provides some important insights into the development of humans over the past few million years. Articles by such leading scientists as Louis S. B. Leakey, Elwyn L. Simons, George Gaylord Simpson, and John R. Napier explore such exciting areas of research as fossil primates, fossil humans, mechanisms of locomotion (such as walking) in primate and hominid species, the immunology of hominids, the evolution of hunting, and personality development during infancy. The articles contain important relevant data, excellent photographs, and supporting reference lists.

Wendt, Herbert. *In Search of Adam*. Translated by James Cleugh. Boston: Houghton Mifflin, 1956.

Wendt's work is another survey of human evolution as uncovered by archaeological research. He concentrates upon the scientists, controversies, and debates involved with the question of humankind's origins. He examines data from animal and human fossils to construct the evolutionary linkages between humankind and beast. Furthermore, he maintains that earth's past had been riddled with catastrophes. Part 4, "Adam Unmasked," concentrates upon our current views of humankind's evolution.

Wenke, Robert J. *Patterns in Prehistory: Humankind's First Three Million Years*. New York: Oxford University Press, 1990.

Wenke's book is an important survey of recent archaeological and anthropological research into the several human species that have existed during the past several million years. Wenke discusses the physical attributes of each species, as well as these species' estimated social and cultural lifestyles and their intelligence. He cites a wealth of data, including major discoveries and theories of leading human evolutionary scientists of the 1970's and 1980's.

Willis, Delta. *The Hominid Gang: Behind the Scenes in the Search for Human Origins*. Introduction by Stephen Jay Gould. New York: Viking Penguin, 1989.
Willis' attractive book is a history of human archaeological research during the twentieth century, with special emphasis upon the extraordinary works of Louis, Mary, and Richard Leakey. Willis highlights this history by interviewing prominent anthropologists, archaeologists, and evolutionary scientists, including Richard Leakey and Stephen Jay Gould. Current and past researches are described in exciting detail. Chapter 1, "The Past and Mr. Leakey," chapter 4, "Handy Man," and chapter 9, "*Homo erectus*," highlight the careers of Louis and Mary Leakey.

Young, Louise B., ed. *Evolution of Man*. New York: Oxford University Press, 1970.
This lengthy collection of essays, quotations, and concepts illustrates the concept of evolution, the mechanisms of evolution, humanity's past, the future of human civilization, and gross misconceptions of evolution, tapping a variety of sources ranging from scientists to tyrants. Topics include the facts supporting evolution, progress versus purpose in evolution, the history of human civilization, humanity's control of the environment, the ability of science to contribute to an improved world, and a discussion asking, "Is Civilization Retarding Man's Evolution?" Contributors include Charles Darwin, Jean Baptiste de Lamarck, Julian Huxley, Alfred Russel Wallace, Friedrich Nietzsche, J. B. S. Haldane, Louis Leakey, P. B. Medawar, Rene Dubos, Loren Eiseley, Ashley Montagu, Theodosius Dobzhansky, and even Adolf Hitler.

Yunis, Jorge J., ed. *Molecular Structure of Human Chromosomes*. New York: Academic Press, 1977.
In this book, Yunis and colleagues present several scientific papers that present compelling molecular evidence that demonstrates the evolutionary relatedness between humans and other primate species. Major topics include comparative DNA studies between humans and apes, chromosomes, repetitive DNA patterns, the structure of chromatin, of heterochromatin, and of euchromatin, somatic cell genetics, and RNA/DNA hybridization studies. Contributors to this volume include Yunis, Ann M. Willey, Michael Tsai, P. L. Pearson, K. W. Jones, G. F. Bohr, Bernard Dutrillaux, and Dale M. Steffensen.

Zimmerman, Burke K. *Biofuture: Confronting the Genetic Era*. New York: Plenum Press, 1984.
Zimmerman's book is a scientific and philosophical treatment of the controversy over genetic engineering (cloning). He begins with an overview of genetics,

including clear descriptions of DNA, genes, and the thousands of essential cellular enzymes/proteins encoded by genes. He describes the famous 1960's cloning experiments by J. B. Gurdon and colleagues using frogs and continues with more recent advances in gene-splicing and cloning. He explains the important applications of cloning to experimental medicine, vaccines against deadly diseases, and human gene therapy to prevent or treat inherited genetic defects (such as Down's syndrome and Tay-Sachs disease). Yet, he also debates the ethics of cloning, public fears, and corporate profit motives in biotechnological research. He asks, "Who decides?" in this controversial new era of human evolution.

Chapter 13
AWAKENING: THE SEARCH FOR EXTRATERRESTRIAL INTELLIGENCE

Ashpole, Edward. *The Search for Extraterrestrial Intelligence*. London: Blandford, 1989.

This book is a serious approach to SETI, the Search for Extra-Terrestrial Intelligence, a scientific attempt to monitor and detect the existence of alien civilizations located elsewhere in the universe. Ashpole describes the history of SETI research, the search methodologies, and famous experiments such as the original Project Ozma search by astronomer Frank Drake in the early 1960's at Green Bank, West Virginia. Ashpole also discusses radio astronomy (SETI's primary tool), identification of suitable stars having planetary systems, and searches for alien space probes and alien artifacts possibly left on earth during ancient visits.

Asimov, Isaac. *Extraterrestrial Civilizations*. New York: Crown, 1979.

Biochemist and writer Asimov presents the reader with a simple, enjoyable examination of life on the earth and other planets and of methods of communicating with intelligent life elsewhere in the universe. Chapter 1, "The Earth," looks at animal evolution, primate intelligence, and the development of civilization by humans. Chapter 3, "The Inner Solar System," describes *Mariner* and *Viking* space probe studies of Mars, Venus, and Mercury. Chapter 4, "The Outer Solar System," uses *Pioneer* and *Voyager* data to study the planetary chemistry of Jupiter, Saturn, and their satellites. Chapter 13, "Messages," looks at various mechanisms of communication with extraterrestrials from other solar systems.

Baird, John C. *The Inner Limits of Outer Space*. Hanover, N.H.: University Press of New England for Dartmouth College, 1987.

Baird's work is an encompassing study of radio telescopes and SETI, means of communicating with extraterrestrials, future space colonies, Unidentified Flying Objects, and humanity's own lack of intelligence in dealing with possible extraterrestrials. Chapters 7, "Intelligence and Value Orientation," and 11, "Decoding the Structure of Language and Thought," raise important questions concerning our approach to communicating with intelligent extraterrestrial civilizations. Chapter 12, "Learning What Is Meant," is aimed at studying ways to bridge the language barrier between two completely alien species who are attempting to talk to each other.

Berendzen, Richard, ed. *Life Beyond Earth and the Mind of Man*. Washington, D.C.: National Aeronautics and Space Administration, 1973.

Based upon a SETI symposium held at Boston University in November, 1972, this short work is a collection of essays by participating symposium scientists on the discovery and nature of extraterrestrial civilizations. Articles by Berendzen, Philip Morrison, Ashley Montagu, Carl Sagan, and George Wald speculate upon the possibilities of life existing elsewhere in the universe, techniques for communicating with intelligent extraterrestrial civilizations, and the need for humanity's contact with extraterrestrials.

Berlitz, Charles, and William L. Moore. *The Roswell Incident*. New York: Grosset & Dunlap, 1980.
In this speculative work, author/explorer/linguist Berlitz and colleague Moore examine one of the most controversial Unidentified Flying Object (UFO) cases ever reported: the supposed crash of an extraterrestrial spacecraft near Roswell, New Mexico, in 1947. Berlitz and Moore examine available evidence and interview numerous individuals involved in the incident, which was explained away by the U.S. Air Force as involving merely a weather balloon. Berlitz and Moore argue, as do many UFO advocates, that the spacecraft crash is the correct explanation and that the Air Force has attempted a cover-up of the crash and the capture of live extraterrestrials.

Billingham, John, ed. *Life in the Universe*. Cambridge, Mass.: MIT Press, 1981.
Billingham's work is a synopsis of a SETI symposium attended by leading scientists in the field and includes position papers, scientific essays, and panel discussions between the participants. Harold P. Klein and Sherwood Chang discuss the origin and evolution of life on earth. Lynn Margulis and James Lovelock, the codevelopers of the controversial Gaia hypothesis, discuss "Atmospheres and Evolution." Ronald Bracewell, Frank Drake, Kenneth Jones, and Bernard Oliver speculate upon the detectability of advanced technological civilizations elsewhere in the universe. Philip Morrison provides concluding remarks to the symposium on this most important topic.

Bracewell, Ronald N. *The Galactic Club: Intelligent Life in Outer Space*. Stanford, Calif.: Stanford University Alumni Association, 1974.
Bracewell's book describes the search for extraterrestrials in general and clear-cut terms. Chapter 1, "Are We Alone?" and chapter 3, "Is Man Unique?" debate the probabilities of intelligent life existing elsewhere in an infinite universe. Chapter 4, "Should We Talk to Them? And How?" discusses the difficulties involved in undertaking extraterrestrial communication. Chapter 9, "What Do You Say to a Non-Human?" examines possible message contents. Chapter 15, "Is Interstellar Travel Possible?" discusses scientific progress in interstellar propulsion vehicles. Impressive diagrams and humorous cartoons accompany each chapter.

Breuer, Reinhard. *Contact with the Stars: The Search for Extraterrestrial Life*. Translated by Cecilia Paine-Gaposchkin and Mark Lowery. San Francisco: W. H. Freeman, 1982.

This scientific, but very understandable, discussion of SETI research provides thorough coverage of the subject. Part 1, "From Molecules to Man," includes information on the origin and evolution of our planetary system, the dynamics of the earth's atmosphere, minimum conditions for life on earth, and alternate evolutionary pathways. Part 2, "Interstellar Exchange of Information," describes the proposed interstellar language called LINCOS, the Frank Drake cosmic message from the Arecibo Observatory in 1974, and the consequences of human contact with extraterrestrials. Part 3, "The Conquest of Space," discusses interstellar spacecraft propulsion systems, black holes, and UFOs.

Cameron, A. G. W., ed. *Interstellar Communication: A Collection of Reprints and Original Contributions*. New York: W. A. Benjamin, 1963.

Cameron, a scientist with the Goddard Space Research Center, has assembled a stellar list of contributors for this early examination of contacting extraterrestrial civilizations. The essays in this volume, some of which are reprints, include "Chemical Evolution" (essay five) by Melvin Calvin, "The Search for Artificial Stellar Sources of Infrared Radiation" (essay eleven) by Freeman Dyson, "How Can We Detect Radio Transmissions from Distant Planetary Systems?" (essay sixteen) by Frank Drake, and "The Problem of Transmission in Interstellar Communication" (essay twenty- one) by Sushu Wang. Other contributors include Cameron, Philip Morrison, Bernard M. Oliver, Edward Purcell, and I. S. Schlovskii.

Christian, James L., ed. *ExtraTerrestrial Intelligence: The First Encounter*. Buffalo, N.Y.: Prometheus Books, 1976.

Christian's work is a captivating collection of philosophical and summary scientific essays by a variety of scientists, thinkers, authors, and even actors. The book begins with a discussion of astronomer Frank Drake's famous equation for estimating the number of intelligent civilizations within the Milky Way galaxy. George Abell presents "The Search for Life Beyond Earth: A Scientific Update." Leonard Nimoy of "Star Trek" provides "Conversations with Mr. Spock." James Royce debates the possibilities of "Consciousness in the Cosmos." From a human perspective, Ronald Smith discusses "The Abdication of Human Intelligence," while Michael Tooley questions "Would ETIs Be Persons?"

Crowe, Michael J. *The Extraterrestrial Life Debate, 1750-1900: The Idea of a Plurality of Worlds from Kant to Lowell*. New York: Cambridge University Press, 1986.

This lengthy historical and philosophical treatise, written by a University of Notre Dame scholar, is an account of the eighteenth and nineteenth century debates over the existence of other worlds and life upon those worlds. Chapter 1, "The Plurality of Worlds Debate Before 1750: A Background Survey," includes the ideas of Copernicus, Bruno, Fontenelle, and Newton. Chapter 2, "Astronomers and Extraterrestrials," describes the views of major early astronomers such as Sir William Herschel, Laplace, and Bode. Chapter 3, "Extraterrestrials and the Enlightenment," probes public views in the United States, Great Britain, and France on the plurality of worlds. Chapter 9, "Religious and Scientific Discussions," examines religious debates over extraterrestrials from 1860 to 1900.

Davoust, Immanuel. *The Cosmic Waterhole.* Translated by Barbara Jachowicz. Cambridge, Mass.: MIT Press, 1991.
Davoust's extremely thorough presentation of SETI research presents both scientific data and thought-provoking speculation on the existence of life elsewhere. Chapter 6 summarizes "The Search for Planetary Systems." Chapters 7 and 8 speculate upon the possible demographies and sociologies of extraterrestrial civilizations. Chapters 10, "Strategies of Searching for Extraterrestrial Civilizations," and 11, "The Search for Signals of Artificial Origin," look at radio astronomical SETI searches. Chapter 12, "The Debate Within Society," examines our sociological ability to deal with extraterrestrials if they exist.

Dick, Stephen Jay. *Plurality of Worlds: The Origins of the Extraterrestrial Life Debate from Democritus to Kant.* New York: Cambridge University Press, 1982.
Dick, an astronomer with the U.S. Naval Observatory, presents a lively philosophical, historical, and scientific discussion of the centuries-old debate concerning the existence of planets around other stars and the possibility of life existing on other worlds. His work begins with chapter 1, "One World or an Infinity of Worlds: The Greek Tradition." He then proceeds through Roman, Christian, and Islamic views of the universe, eventually culminating in the works of Galileo Galilei, Nikolaus Copernicus, Johannes Kepler, and Bernard le Bovier de Fontenelle.

Edelson, Edward. *The Search for Intelligent Life in the Universe: Who Goes There?* Garden City, N.Y.: Doubleday, 1979.
Edelson approaches a number of important topics in SETI research, including the origins of life, evolution, and communication with extraterrestrials. Chapter 1, "Viking to Mars," presents the results of exobiological experiments aboard the *Viking 1* and 2 Mars landers. Chapter 6, "The Formula," describes the Drake equation and its implications for studying extraterrestrials. Chapters 8, "Travelling," 9, "Listening," and 10, "Messages," elaborate the various mechanisms of contacting extraterrestrial civilizations. Chapter 11, "Strategies," discusses American and Soviet efforts at detecting the existence of extraterrestrials. Chapter 12, "Futures," offers humanity two alternatives: "space or stagnation."

Feinberg, Gerald, and Robert Shapiro. *Life Beyond Earth: The Intelligent Earthling's Guide to Life in the Universe*. New York: William Morrow, 1980.
This speculative scientific discussion of the possibilities of life existing elsewhere in the universe and how to find it includes very relevant astronomical and biological data from origin-of-life research experiments on earth to interplanetary space probes (such as the *Viking* missions to Mars). Chapters 2, "The Architecture of Life," 3, "Earth's Biosphere and What It Eats," 6, "What Is Life?" and 7, "Conditions for Life," all examine the nature of living organisms and the necessary environmental constraints for the origin, evolution, and sustenance of life. Chapters 13, "The Creatures of Elsewhere," and 14, "A Universe Full of Life and How to Find It," tackle the question of how life might exist elsewhere under drastically different conditions and discuss search strategies for detecting life on other worlds.

Fontenelle, Bernard le Bovier de. *Conversations on the Plurality of Worlds*. Translated by H. A. Hargreaves. Berkeley: University of California Press, 1990.
Originally written in 1686, this daring work by a French philosopher/scientist supported the scientific views of Galileo Galilei and Nikolaus Copernicus in spite of the then-totalitarian Catholic church. Fontenelle proposes that the earth is but one of several worlds within the universe and that life exists on other worlds. He presents his arguments within the context of five evening discussion sessions with friends and improvised characters. The work is very short and clear.

Fuller, John G. *Incident at Exeter*. New York: G. P. Putnam's Sons, 1966.
Fuller presents an objective account of UFO sightings and encounters in the Exeter, New Hampshire, area in 1965. Initially a skeptic, Fuller interviews dozens of witnesses and examines considerable reliable evidence to conclude that the UFO sightings were probably extraterrestrial spacecraft. He cites U.S. Air Force attempts to conceal or explain away UFO encounters, including smearing of reliable witnesses by Air Force UFO investigators.

Goldsmith, Donald, ed. *The Quest for Extraterrestrial Life: A Book of Readings*. Mill Valley, Calif.: University Science Books, 1980.
Goldsmith's oversized book is a collection of original essays and reprinted scientific articles discussing the origins of life and of SETI. Numerous articles by many leading scientists are presented. Francis Crick and Leslie Orgel present their controversial "Directed Panspermia" hypothesis in essay ten, a hypothesis that maintains that life on earth may have been seeded from outer space. Sidney Fox discusses "Humanoids and Proteinoids" in essay forty. Michael Papagiannis queries "Are We Alone, or Could They Be in the Asteroid Belt?" in essay forty-six. Other contributors include the ancient Greek philosopher Lucretius, Christian Huygens, Voltaire, J. B. S. Haldane, Svante Arrhenius, Fred Hoyle, Geerat Verschuur, Bernard Oliver, Edward Purcell, Carl Sagan, George Gaylord Simpson, Lewis Thomas, and Gerard O'Neill.

Goldsmith, Donald, and Tobias Owen. *The Search for Life in the Universe*. Menlo Park, Calif.: Benjamin/Cummings, 1978.
This comprehensive book traces the quest for life in the universe from the origins of life on earth to attempts at communication with possible extraterrestrial civilizations. Part 1, "Why Do We Search?" describes from a human perspective the importance of discovering life elsewhere and the search for the origins of life on the earth. Part 2, "The Universe at Large," presents the life cycles of stars, the history of galaxies, and the theoretical basis of space-time geometry. Part 3, "Life," looks at "The Origin of Life" (chapter 9) and the evolution of life in "From Molecules to Minds" (chapter 10). Part 5, "The Search for Extraterrestrial Intelligence," queries "Is Earth Unique?" (chapter 17), "The Development of Extraterrestrial Civilizations" (chapter 18), and "How Can We Communicate?" (chapter 19). Other chapters deal with Unidentified Flying Objects (UFOs) such as the ancient astronauts of Erich von Daniken, etc.

Good, Timothy. *Above Top Secret: The Worldwide U.F.O. Coverup*. Foreword by Lord Hill-Norton. New York: William Morrow, 1988.
English UFOlogist Good has compiled an impressive, objective coverage of unexplained UFO encounters throughout the world from the 1940's to the present in this thorough, lengthy volume. He cites more than one hundred well-documented cases of UFO encounters involving credible witnesses. He attacks several foreign governments and the U.S. Air Force for attempting UFO cover-ups in the interest of national security. His writing is objective, avoiding obvious UFO hoaxes and ridiculous accounts of UFO encounters arising in the tabloid media.

Graham-Smith, Francis, and Bernard Lovell. *Pathways to the Universe*. New York: Cambridge University Press, 1988.
In this important work, two leading English astronomers combine to produce a scientific survey of astronomy, cosmology, and the possibility of life existing throughout the universe. They describe the evolution of the universe, the distributions of galaxies throughout space-time, the life cycles of stars, the formation of complex elements within stars, the formation of planetary systems, and the origin and evolution of life on earth. They describe the tools of astronomy, including the science of radio astronomy, which is extensively used in SETI research.

Hart, Michael H., and Ben Zuckerman, eds. *Extraterrestrials: Where Are They?* New York: Pergamon Press, 1982.
Essays by noted biologists, chemists, physicists, and astronomers highlight this important contribution to the study of exobiology, the science of extraterrestrial life. Zuckerman describes mechanisms such as radio astronomy for detecting extraterrestrials in "Searches for Electromagnetic Signals from Extraterrestrial Beings," chapter 2. Freeman Dyson considers the possibilities of finding life elsewhere via space travel in "Interstellar Propulsion Systems," chapter 6. Cyril

Ponnamperuma speculates upon "Primordial Organic Chemistry" in chapter 11, while Michael Hart examines "Atmospheric Evolution, the Drake Equation, and DNA: Sparse Life in an Infinite Universe" in chapter 12. Data tables and strong reference lists support each article.

Jones, H. Spencer. *Life on Other Worlds*. New York: Macmillan, 1940.
This short work, although dated, is an excellent examination of the possibility of life existing on other worlds. Jones compares the environmental conditions on earth to conditions on other planets in order to speculate upon the probabilities of life existing elsewhere within our own solar system. He relies primarily upon comparative planetology, the comparison of other planetary environments to the atmospheric and environmental conditions needed to sustain life on earth. He focuses on the planets Mars and Venus in particular.

McDonough, Thomas R. *The Search for Extraterrestrial Intelligence*. New York: John Wiley & Sons, 1987.
McDonough's best-selling work is a partly serious, partly humorous look at the search for alien civilizations located elsewhere in the universe. He discusses SETI attempts performed hundreds of years ago, public hysteria over Martians and UFOs, and the determined efforts by leading astronomers to detect life elsewhere, principally by analyzing incoming radio signals from outer space. McDonough writes with great simplicity and incorporates interesting photographs and cartoons both to educate and to entertain the reader.

MacGowan, Roger A., and Frederick Ordway III. *Intelligence in the Universe*. Englewood Cliffs, N.J.: Prentice-Hall, 1966.
MacGowan and Ordway's extraordinary book is a discussion of the evolution of the cosmos, the origin and evolution of life, the evolution of intelligent life on earth, and speculations on the nature of extraterrestrial civilizations that most probably are very different from our own civilization. Early chapters are devoted to cosmology, the evolution of planetary systems, and the evolution of life on earth. Chapters 11, "Characteristics of Extrasolar Intelligence," and 12, "Capabilities of Extrasolar Intelligence," speculate upon the nature and technological levels of possible contact civilizations. Chapters 15, "Empirical Evidence of Extrasolar Intelligence," and 16, "Communications with Extrasolar Intelligence," tackle the difficulties of identifying and contacting potential sites of extraterrestrial civilizations.

Macvey, John W. *Alone in the Universe?* New York: Macmillan, 1963.
Macvey's book is primarily a philosophical discussion of the likelihood of life existing on other worlds in our own and other solar systems. He supports his arguments with available scientific data. He talks about the lifetimes and evolution of stars, possible planetary systems around some stars, alternative environmental

conditions on other worlds that would lead to alternate evolutionary pathways if life arose, and conditions on solar system planets such as Mars, Venus, and Saturn.

Ovenden, Michael W. *Life in the Universe: A Scientific Discussion.* Garden City, N.Y.: Doubleday, 1962.
This short, clear book is a discussion of the possibility of life existing elsewhere in our solar system and throughout the universe, based upon earth as a model for a suitable life environment. Ovenden describes the fundamental conditions on earth that are primarily responsible for the occurrence and maintenance of life. He discusses the evolution of the solar system, environmental conditions on other planets and satellites such as Mars, possible planetary systems around other stars, and the origin and adaptability of life.

Papagiannis, Michael D., ed. *The Search for Extraterrestrial Life: Recent Developments.* Boston: D. Riedel, 1985.
This lengthy, detailed work is a compilation of SETI essays by leading astronomers and exobiologists for the International Astronomical Union. Contributors include Linus Kafatos, Jill Tarter, Robert S. Dixon, Cyril Ponnamperuma, Leslie Orgel, John H. Wolf, Eric M. Jones, John A. Ball, George Gatewood, and Harlan J. Smith. Breakthroughs in origin-of-life research, space probes (such as *Voyager 1* and 2) to the outer planets, the detection of possible planetary systems around nearby stars, and SETI search experiments using radio telescopes are summarized in these informative reports.

____________. *Strategies for the Search for Life in the Universe.* Boston: D. Riedel, 1980.
Papagiannis' work is a collection of essays and panel discussions by leading SETI researchers during a special session at the August, 1979, meeting of the International Astronomical Union in Montreal. Participants at the SETI session include Papagiannis, Frank Drake, J. L. Greenstein, Michael H. Hart, Philip Morrison, Bernard M. Oliver, Jill Tarter, and Ben Zuckerman. They discuss a variety of relevant topics such as microwave and radio searches for artificial extraterrestrial signals, radio leakage, the Drake equation, and starships and stellar travel. Papagiannis provides concluding remarks.

Pirsig, Robert M. *Zen and the Art of Motorcycle Maintenance: An Inquiry into Values.* New York: Bantam Books, 1974.
While Pirsig's book does not deal with extraterrestrial life or evolution, it does address the human condition on earth. Initially rejected by more than one hundred publishers, Pirsig's philosophical best-seller is an autobiographical account of a man's search for meaning and quality in life. He stresses the need for people to open their minds and to think differently about the world, to live with purpose

and accomplishment. He stresses important concepts such as "lateral thinking" and "selfless climbing" in living in touch with other people and the universe. Required reading.

Pittendrigh, Colin S., Wolf Vishniac, and J. P. T. Pearman, eds. *Biology and the Exploration of Mars: Report of the Study Held Under the Auspices of the Space Science Board, National Academy of Sciences, National Research Council, 1964-1965*. National Research Council Publication 1296. Washington, D.C.: National Academy of Sciences, 1966.
In this comprehensive work, many notable astronomers, physicists, and biochemists supply scientific assessments of the possibility of life on the planet Mars and methods for studying that planet. N. H. Horowitz, Joshua Lederberg, Stanley Miller, Carl Sagan, and Harold Urey discuss prebiotic chemistry and the origin of life on earth. Sidney Fox, G. C. Pimentel, Sagan, and Vishniac speculate upon the nature of extraterrestrial life and intelligence. E. C. Levinthal, Miller, and Sagan suggest various approaches for studying Mars. A. H. Brown, Horowitz, and Levinthal discuss possible manned explorations of the planet's surface.

Ponnamperuma, Cyril, and A. G. W. Cameron, eds. *Interstellar Communication: Scientific Perspectives*. Boston: Houghton Mifflin, 1974.
Exobiologists Ponnamperuma and Cameron have assembled a stunning collection of scientific papers by leading experts in the study of extraterrestrial life. The contributed articles to this important volume concentrate upon methods and approaches to contacting and communicating with intelligent extraterrestrial civilizations. Contributors include Michael Arbib, Ronald Bracewell, Cameron, Frank Drake, John McCarthy, Philip Morrison, Bernard Oliver, Ponnamperuma, and Carl Sagan. The first article, by Sagan, describes the Drake equation and the constraints that it applies to the possibilities of life elsewhere in the universe.

Regis, Edward, Jr., ed. *Extraterrestrials: Science and Alien Intelligence*. New York: Cambridge University Press, 1985.
This collection of scientific and philosophical essays by leading scientists is an objective approach to the consideration of extraterrestrial life and scientific methods of contacting intelligent extraterrestrials. Contributors include Ernst Mayr, David M. Raup, Frank J. Tippler, Carl Sagan, Jill Tarter, and Hans Freudenthal. Raup discusses "ETIs Without Intelligence," whereas Tippler argues that "ETs Don't Exist." Freudenthal presents "Excerpts from LINCOS: Design of a Language for Cosmic Intercourse." An appendix to this volume summarizes all international astronomical SETI observing programs from 1959 through 1984.

Ridpath, Ian. *Messages from the Stars: Communication and Contact with Extraterrestrial Life*. New York: Harper & Row, 1978.
Ridpath's book is an exciting discussion of the possibilities of intelligent life

existing elsewhere in the universe. Chapter 1 asks "Are We Alone?" Chapter 6, "Our First Ship to the Stars," describes the robot space probes *Pioneer 10*, *Pioneer 11*, *Voyager 1*, and *Voyager 2*, all of which will have left our solar system by the early 1990's. Chapter 14, "The Siberian Spaceship Explosion," discusses the 1908 Tunguska event, in which an object (comet? spaceship?) struck the earth. Other chapters deal with possible visitations to earth by ancient astronauts.

__________. *Worlds Beyond: A Report on the Search for Life in Space*. New York: Harper & Row, 1975.
Extensive beautiful photographs and diagrams highlight Ridpath's extraordinary discussion of the evolution of life on earth and the search for life elsewhere in the universe. Ridpath describes important origin-of-life experiments such as the famous Stanley Miller-Harold Urey experiment of the 1950's. Chapter 6 looks at the possibilities of "Life in the Solar System." Chapters 8, "Travel to the Stars," and 9, "Probes from the Skies," examine modern techniques for detecting life elsewhere, including robot space probes and SETI search experiments using radio telescopes.

Rood, Robert T., and James S. Trefiltart. *Are We Alone? The Possibility of Extraterrestrial Civilizations*. New York: Charles Scribner's Sons, 1981.
This informative book devoted to the evolution of life in the cosmos and to human attempts at contacting extraterrestrial life-forms utilizes simple explanations of complex physical and astronomical concepts. The authors describe the lives of stars, star and planetary formation, planetary atmosphere formation, the origin and evolution of life on earth, the evolution of intelligence and intelligent civilizations in the universe, the Drake equation, human and biological chauvinisms concerning the nature of life in the universe, and SETI experiments. Chapter 11 appropriately asks, "Is There Intelligent Life on Earth?"

Sable, Martin H. *Exobiology: A Research Guide*. Brighton, Mich.: Green Oak Press, 1978.
This work, written by a University of Wisconsin-Milwaukee professor, is a bibliography of books, scientific journal articles, government agency reviews, magazine articles, and newspaper articles devoted to the existence of life on other planets. Nearly four thousand works are cited by title, author, and publisher, followed by an alphabetical listing of authors. There is no commentary on book/article contents. Titles include the existence of intelligent life on and off earth, the search for life on other worlds, humanoid contacts on earth, teleportations, and Unidentified Flying Objects. The late UFOlogist Dr. J. Allen Hynek provides a "cover letter" for the book.

Sagan, Carl, ed. *CETI: Communication with Extraterrestrial Intelligence.* Cambridge, Mass.: MIT Press, 1971.
This fascinating little book is a transcript of the first international conference devoted to the search for extraterrestrials, held in Soviet Armenia in 1970. Lively, clear discussions by many of the world's top astronomers, physicists, chemists, biologists, and anthropologists illustrate the problems and the importance of SETI research. Conference participants include Sagan, Philip Morrison, Francis Crick, David Hubel, Nikolai Kardashev, V. A. Ambartsumian, and Frank Drake. The book is organized around the Drake equation, an estimate of the number of extraterrestrial civilizations in the Milky Way galaxy.

Schlovskii, I. S., and Carl Sagan. *Intelligent Life in the Universe.* San Francisco: Holden Day, 1966.
Two principal pioneers in SETI research, one a Soviet, the other an American, collaborated on this classic work on the subject of communicating with other worlds. Together, Schlovskii and Sagan delineate the parameters of detecting alien civilizations and the possible techniques of communicating with them. They examine possible star systems for study, evolutionary cosmology, the Drake equation, and the use of radio astronomy and automated space probes to detect extraterrestrial life. Excellent photographs and illustrations support the outstanding text of this must reading.

Sullivan, Walter. *We Are Not Alone: The Search for Intelligent Life on Other Worlds.* New York: McGraw-Hill, 1964, 1966.
This lengthy scientific discussion of SETI is an early examination of the subject. Chapter 5, "The Solar System: Exception or Rule?" looks at the possibilities of planetary systems forming around other stars. Chapter 7, "Where to Look," considers candidate star systems for scrutiny in the search for life. Chapter 12, "Is There Life on Mars?" represents a pre-*Viking* examination of the possibility of life on that planet. Chapter 14, "Project Ozma," details astronomer Frank Drake's pioneering SETI experiment. Chapter 17, "Is There Intelligent Life on Earth?" asks a very important question.

Swift, David W. *SETI Pioneers.* Tucson: University of Arizona Press, 1990.
This interesting work, written by University of Hawaii sociologist Swift, is a compilation of interviews with seventeen leading SETI research scientists, including Carl Sagan, Frank Drake, Bernard Oliver, Nikolai Kardashev, Philip Morrison, Melvin Calvin, and Freeman Dyson. Discussed topics include the history of SETI research experiments from Drake's Project Ozma to NASA's Project Cyclops to current efforts, problems with search strategies (quite literally a needle in a haystack), and future SETI research.

INDEX

INDEX

INDEX

INDEX

INDEX

INDEX